Telecommunications, Networking and Internet Glossary

George S. Machovec
Colorado Alliance of Research Libraries

LITA Monographs 4

Library and Information Technology Association
a division of the
American Library Association
Chicago and London 1993

LITA Monograph Series
No. 1 *Library and Information Technology Standards.*
Edited by Michael Gorman
No. 2 *Access to Information: Materials, Technologies, and Services for Print-Impaired Readers*
By Tom McNulty and Dawn M. Suvino
No. 3 *Internet Connections: A Librarians Guide to Dial-Up Access and Use*
By Mary E. Engle, Marilyn Lutz, William W. Jones, Jr., Genevieve Engel

Composition by George Machovec, Colorado Alliance of Research Libraries, using Ami Pro 3.0. Camera-ready pages output on a HP LaserJet 4 printer at 600 dpi.

The paper used in this publication meets the minimum requirements of the American National Standard for Information Sciences-Permanence of Paper for Printed Library Materials, ANSI Z39.48-1984. ∞

ISBN 0-8389-7697-2

Printed in the United States of America

97 96 95 94 5 4 3 2

Contents

Acknowledgments

I wish to thank, Alan Charnes, Executive Director, Colorado Alliance of Research Libraries (CARL), and Sharon Andersen, Executive Assistant, CARL, for their support on this project. They have made my move to Denver, CO in January 1993 much easier with their help, advice and encouragement.

Ann Hope, LITA Monographs Editor, has done an excellent job in reviewing the manuscript and in suggesting changes.

I would also like to thank the libraries at Arizona State University and the University of Denver for the use of their collections. Both institutions have excellent collections and the reference staff at each site has been very accommodating in providing special borrowing privileges.

I would also like to express my gratitude to my wife, Nina, and my two children, Rachel and Alexander, for their understanding in the time required to put together a work such as this.

Preface

Networking and telecommunications issues are inextricably linked with all phases of online information retrieval and library automation. One of the major problems facing information managers and librarians is that the terminology in these areas is highly specialized and is growing at a phenomenal pace. The networking and telecommunications fields are immense and it is impossible to provide a complete list of all technical terms or acronyms. However, this is an attempt to provide brief non-technical definitions of terms and acronyms which frequently appear in library automation and computer literature.

This is a greatly revised and expanded edition of the *Telecommunications and Networking Glossary* (LITA Guides 3) which was published in 1990. The name has been changed to reflect a new major emphasis -- namely terms relating to the Internet and NREN. The Glossary has more than doubled in size and many of the additions are items relating to the Internet, NREN, TCP/IP, OSI and other similar topics. Many of the new terms in this volume did not even exist when the 1990 Glossary was compiled!

Since the purpose of this work is to focus on networking and telecommunications, computer or library automation terminology which do not directly relate to the intended scope are not included. General terms concerning computer technology, library automation and information technology may be found in a number of other dictionaries.

One of the major difficulties in compiling a dictionary of this nature is that many of the terms widely used in the field are for proprietary products and services. Many dictionaries ignore this type of terminology to the detriment of the reader who needs these definitions. An attempt has been made to include both proprietary and non-proprietary terminology when it is considered that a word or phrase is frequently used.

Another major difficulty in compiling glossaries is that one always seems to miss "important" key words, at least according to someone else's definition. Give practically any dictionary or glossary to a group of readers and they will all discover some of their favorite terms missing. Although it is virtually impossible to get around this dilemma, especially in a brief glossary, the terms included in this list have been checked against some of the works listed in the bibliography at the end of the work.

It is realized that entire works and fields of study exist on most of the terms in this glossary. In order to obtain more background information on the terms a highly selective bibliography of recent books, journal articles and conference proceedings on networking and telecommunications is provided to offer further readings on each of the definitions provided.

It is hoped that this guide will serve as a useful handbook to librarians, information managers and students involved with networking, telecommunications and the Internet.

George S. Machovec
September 1993

Telecommunications, Networking and Internet Glossary

AARE An abbreviation for "A-Associate Response."

AARNet The Australian Academic and Research Network. The major research network in Australia founded in 1990 which offers networking services for research and academic institutions.

AARQ An abbreviation for "A-Associate Request."

ABBREVIATED ADDRESS CALLING A subset of address characters used to initiate a call from a terminal. The network expands the abbreviation to a full address. In electronic mail systems these abbreviations are often called nicknames.

ABBREVIATED DIALING SERVICE A technique used in packet switching networks whereby nonnumeric codes are available to represent groups of digits or prefixes. This technique is used in speed dialing by the public telephone service.

ABM Asynchronous Balanced Mode. One of three HDLC operating modes. It provides a two-way simultaneous data link with the two communicating stations, referred to as "combined stations," having equal (balanced) responsibility for link management. *see also* High Level Data Link Control.

ABORT TIMER A device which monitors the receive end of a data communications circuit. If no data are sent within a preset time the connection is terminated.

ABRT An abbreviation for "abort."

ABSTRACT SYNTAX (AS) A data structure which is often an application layer protocol data unit (PDU).

ABSTRACT SYNTAX NOTATION ONE *see* ASN.1.

AC *see* Access Control.

AC An abbreviation for "alternating current."

AC SIGNAL A signal in which the direction of the current is switched on a regular frequency.

ACADEMNET This network offers connections from the Institute for Automated Systems in Moscow to institutes in Russia and some former republics.

ACCENTUATED CONTRAST A technique used with telefacsimile machines to darken those parts of an image with a certain threshold of opacity and to make white those parts of an image that fall under the threshold.

ACCESS CHANNELS Used to connect a local subscriber's equipment to a long distance network (often phone lines).

ACCESS CHARGE A cost which local telephone companies charge to use their phone network to originate long distance traffic. These fees are approved by the FCC and are meant to balance a loss in revenue to local phone services because of the loss of cross subsidies resulting from the breakup of the Bell system. This is also known as carrier common line charge (CCLC).

ACCESS CODE A prefix or password to an address that allows a user to obtain access to a particular service.

ACCESS CONTROL A field used in the IEEE 802.5 link-level protocol. It is also used more generally to refer to any security methods used to limit user access to a network or computer system.

ACCESS METHOD The means by which devices may gain access to a network, such as a LAN, in order to transmit data. Common methods include token-passing and CSMA/CD.

ACCESS PATH A path providing communication between two devices. The path may not always be direct and may go through intermediate nodes on a network.

ACCESS POINT The actual point of connection, or the interface, to a network.

ACCESS PROTOCOL The set of rules that LAN workstations must use to send data over a shared network media. Examples would be CSMA and token passing. This is a synonym to Media Access Control (MAC).

ACCESS RIGHTS Permission to access a computer system or perform some combination of using, reading, or writing to a computer file. On a network users may have different access rights.

AccessNB The New Brunswick/Prince Edward Island Education Computer Network was founded in the 1960's. It is a mid-level CAnet network and serves New Brunswick and Prince Edward Island in Canada.

ACCUMASTER CONSOLIDATED WORKSTATION (ACW) An MS-DOS based system that displays a consolidated network view by providing separate windowing sessions to various AT&T EMSs (Element Management System).

ACCUMASTER INTEGRATOR AT&T's UNIX-based system capable of automatically uploading information from other element management systems via the NMP (Network Management Protocol) interface.

ACD *see* Automatic Call Director.

ACID An abbreviation for "atomicity, consistency, isolation and durability."

ACK An acknowledge character. This control character is transmitted by a station as a positive response to the station with which it has connected.

ACKNOWLEDGMENT SIGNAL UNIT (ACU) A block of signal information which indicates whether the other data in a block were properly received.

ACONet Akademisches Computer Netz (Academic Computer Network). Founded in 1982, ACONet provides networking for research and higher education institutions in Austria.

ACOUSTIC COUPLER A device into which one puts a telephone headset to receive and transmit data. The unit converts digital signals into analog signals which allows the data to be transmitted over analog public phone lines. These are outmoded devices and are now rarely sold although many are still in use with microcomputers and terminals in homes.

ACS *see* Advanced Communication Systems.

ACSE Association Control Service Element. An ISO protocol providing a set of common association control services to other application service elements such as to establish, terminate or abort services. It is a grouping of application services to establish and release associations. *see also* Association.

ACSS Advanced Communications Support Systems. A broadband communications backbone developed by Sytek, Inc.

ACTIVITY In an OSI compliant network this refers to an identified operation at the session layer such as the transmission of a message.

ACTUAL FINAL ROUTE The path taken by a call or packet of information over an international switched telephone network. This can be contrasted with the theoretical final route.

ACU *see* Acknowledgment Signal Unit.

ACUTA The Association of College and University Telecommunications Administrators.

ADAPTER A unit which interfaces a data terminal to a communications channel. It may do speed or code conversion to make the communications device compatible with the channel.

ADAPTIVE CHANNEL ALLOCATION A technique used in time division multiplexing in which the data capacity of a channel is determined as a consequence of demand.

ADAPTIVE DIFFERENTIAL PULSE CODE MODULATION (ADPCM) A technique whereby sound can be digitized through examining successive analog waveforms and transmitting the changes in these waveforms rather than transmitting the actual waveforms themselves. Due to the reduction in the amount of data needed to be transferred, an effective data transfer rate of 32 Kbps can be achieved by this technique. *see also* Pulse Code Modulation.

ADAPTIVE EQUALIZER A device used in analog communication circuits which counteracts delay distortion.

ADCP An alternate abbreviation for "advanced data communication protocol." *see* ADCCP.

ADCCP Advanced Data Communication Control Procedure. An ANSI X3.66 bit-oriented protocol, which may be compared with the ISO HDLC protocol. It specifies error control protocols for data transmission.

ADDRESS The origination source of or destination point where data will be sent. The address can be coded in many different ways depending on the type of telecommunications or networking protocols being used.

ADDRESS FIELD A field of information in a frame of data which identifies the unique address for the data being sent.

ADDRESS RESOLUTION PROTOCOL (ARP) A LAN protocol used by the military for resolving local SNPA (subnet point of attachment) connectivity problems.

ADMD Administrative Management Domain. A Message Handling Service (MHS) domain managed by a network administrator.

ADPCM - *see* Adaptive Differential Pulse Code Modulation.

ADR An abbreviation for "address."

ADVANCED COMMUNICATION SYSTEMS (ACS) A system developed by AT&T which allows incompatible terminals to communicate with each other. ACS was later called Advanced Information Systems/Net 1.

ADVANCED COMMUNICATIONS SUPPORT SYSTEMS *see* ACSS.

ADVANCED DATA COMMUNICATIONS CONTROL PROCEDURES *see* ADCCP.

ADVANCED NETWORK & SERVICES *see* ANS.

ADVANCED PROGRAM TO PROGRAM COMMUNICATION *see* APPC.

ADVANCED RESEARCH PROJECTS AGENCY *see* ARPA.

ADVANCED RESEARCH PROJECTS AGENCY NETWORK *see* ARPANET.

AE *see* Application Entity.

AERONAUTICAL TELECOMMUNICATIONS NETWORK (ATN) A collection of networks for the exchange of aeronautical information between civil aviation authorities and airline organizations, including the aircraft.

AES *see* Aircraft Earth Station.

AFI *see* Authority and Format Identifier.

A/G Air-to-Ground. Radio communication channels used by stations (on either the ground or in aircraft) in the aeronautical mobile service.

AFIPS - *see* American Federation of Information Processing Societies.

AFS Andrew File System. A suite of protocols which allows the ability to use files on other network computers as if they were local. So rather than downloading a file to a local computer, the user can read it, write it and edit it on the remote computer. This product is not yet in widespread use but a commercial version is available from Transarc. It offers a newer alternative to the NFS (network file system). *see also* NFS.

AGC *see* Automatic Gain Control.

AGFNET Arbeitsgemeinschaft der Grossforschungseinrichtungen (The Association of National Research Centers of the Federal Republic of Germany). This network provides backbone communications services in Germany.

AGGREGATE INPUT The total of all data rates from computer ports connected to a concentrator or multiplexer.

AIR TRAFFIC CONTROL (ATC) A government organization which provides control service for aircraft as well as offering weather, Notams and other advisory information for aircraft.

AIRCRAFT EARTH STATION (AES) A satellite station used on commercial aircraft for receiving satellite signals used for air-to-ground voice and data communications.

AIX A form of the UNIX operating system that operates on IBM UNIX workstations.

ALA An abbreviation for the "American Library Association."

ALANET The electronic mail and information service of the American Library Association (ALA). This service was discontinued.

ALPHAGEOMETRIC CODING A technique for encoding, transmitting and displaying of graphics information for a videotex system.

ALPHAMOSAIC CODING A technique for encoding, transmitting and displaying of text and graphics to be used in a videotex system.

ALT An alternate USENET category for newsgroups relating to "alternative" topics.

ALTERNATE MARK INVERSION (AMI) A technique used in data communications in which the mark condition of a coded signal is alternately represented by a positive and negative voltage on the line.

ALTERNATE PATH ROUTING In networking, this is an access path other than the normal or basic access path which must be used because of heavy traffic or other problems in the network.

ALTERNATIVE OPERATOR SERVICE (AOS) An organization that provides operator services but is other than one of the Bell operating companies (BOCs) or AT&T.

ALTERNET AlterNet is run by UUNET and acts as a public TCP/IP network access point to the Internet for any organization or individual. Dial-up and leased line access are available to NSFNET mid-level regional networks including PSInet and CIX and others. *see also* UUNET.

AM *see* Amplitude Modulation.

AMERICAN FEDERATION OF INFORMATION PROCESSING SOCIETIES (AFIPS) An organization of computer related societies including the Institute of Electrical and Electronic Engineers, the Association for Computing Machinery, the American Society for Information Science, Simulation Councils Inc. and others.

AMERICAN NATIONAL STANDARDS INSTITUTE *see* ANSI.

AMERICAN STANDARD CODE FOR INFORMATION INTERCHANGE - *see* ASCII.

AMI *see* Alternate Mark Inversion.

AMPLITUDE MODULATION (AM) A method for modifying a carrier signal, which is a sine wave, in order to transmit information.

ANALOG Refers to the technique of conveying information by modulating (or varying) a wave by frequency, amplitude or phase of the carrier. This should be contrasted with digital.

ANALOG CIRCUIT A channel in a data communications system in which data can have any value between the limits defined by the circuit.

ANALOG SIGNALS Signals that vary continuously by amplitude or frequency. Historically older transmission systems have been analog (e.g. the phone service) although most newer systems are digital in nature. Compare with digital signals.

ANALOG VIDEO A signal used for broadcast television.

ANDREW FILE SYSTEM *see* AFS.

ANI *see* Automatic Number Identification.

ANONYMOUS FTP *see* FTP.

ANS Advanced Network and Services. A U.S. network service provider for the Internet which supports Internet connections for research and educational institutions, government agencies and others. It was the first provider of a T3 (45 Mbps) backbone for public access on the Internet. It was formed in September 1990 by NSF, MERIT, IBM and MCI as a 501 (c) (3) not-for-profit entity, with the stated goal of eventually developing a gigabit network infrastructure for American research and education.

ANS CO+RE Advanced Network and Services (ANS) provides network connections to the Internet with a variety of bandwidths. ANS CO+RE is a wholly owned subsidiary of ANS which provides Internet access to commercial organizations, including other network and information service providers.

ANSI American National Standards Institute. This nonprofit, nongovernmental organization began in 1918 and now consists of over 1,000 professional associations, trade organizations and corporations. It serves as a national repository for voluntary standards and is the representative organization in the United States before the International Standards Organization (ISO).

ANSWERBACK An automatic or manual response from a device that indicates that the correct device has been reached and is operational.

ANTERIOR TECHNOLOGY A communications service provider for BARRnet and PSInet to support Internet access, USENET news distribution and other value added computer communications. It is located in Menlo Park, California.

AOCE *see* Apple Open Collaboration Environment.

AOS *see* Alternative Operator Service.

AOW An abbreviation for "Asia-Oceania Workshop."

AP An abbreviation for "application program."

AP *see* Application Process.

APDU An abbreviation for "application protocol data unit." *see also* Protocol Data Unit.

API Application Program Interface. This is software which allows applications programs to talk to communications software, thus allowing particular applications to be developed separately from communications.

APPC Advanced Program-to-Program Communication. This API is a set of programs developed by IBM to allow its SNA networks to communicate without the need of a mainframe (also called LU6.2). These programs establish the required conditions that enable application programs to send data to each other through the network (at the OSI model's session layer).

APPLE OPEN COLLABORATION ENVIRONMENT (AOCE) A product being developed for Macintosh computers in which all incoming data traffic (e.g. email, fax, voice mail, news service feeds, reminders) will be presented to the user in a single "in box." It will also provide a single directory for services.

APPLICATION Software that is written for a particular function.

APPLICATION ENTITY (AE) The implementation of a group of application service elements (ASE's) and local software, in combination with the Application Process being serviced, in one end system. The purpose of the invocation is to create a logical communications path between one AE and a like AE in a remote system.

APPLICATION LAYER In the OSI model this is the 7th layer (highest) and is responsible for providing application programs for the end user. This layer does not specify what applications will be available to the user but rather what communications services are to be available to a computer system for a number of specific purposes.

APPLICATION PROCESS (AP) A user of the OSI communication services.

APPLICATION PROGRAM INTERFACE *see* API.

APPLICATION SERVICE ELEMENT (ASE) An application layer entity which performs a defined set of services and functions. It includes sets of related services that are grouped together under one standard.

ARCHIE The McGill School of Computer Science Archive Server Listing Service. Archie consists of two software tools: the first keeps track of Internet FTP archive sites in a central server and is updated about once per month; the second tool allows users to query this database to identify who owns what. Archie keeps track of both UNIX and VMS archive sites on the Internet and the software has been loaded at different institutions around the world which are connected to the Internet.

ARCNET Datapoint has developed this low-cost LAN which operates at 2.5 Mbps per second and uses a token passing bus architecture, usually on a coax cable.

AREA CODE A three digit number which identifies geographical areas in the long distance telephone service in the United States.

ARIADNE The Greek communications network offering data communications to academic and research institutions throughout Greece.

ARISTOTE Association de Reseaux Informatique en Systeme Totalement et Tres Elabore (Association of Information Networks in a Completely Open and Very Elaborate System). Founded in 1987, this network represents an association of French research institutions developing a networking technology.

ARL *see* Association of Research Libraries.

ARM *see* Asynchronous Response Mode.

ARnet The Alberta Research Network was founded in January 1990 which serves as an academic and public data communications network in Canada.

ARP *see* Address Resolution Protocol.

ARPA Advanced Research Projects Agency. This U.S. government agency which was part of the U.S. Department of Defense built one of the first inter-networks called ARPANET. *see also* ARPANET.

ARPANET (ADVANCED RESEARCH PROJECT AGENCY NETWORK) A federally operated computer packet switching network which was originally developed by the Defense Advanced Research Projects Agency (DARPA) in 1968. In 1975, the operating responsibility was given to the Defense Communications Agency. The network primarily serves the federal government and its contractors. TCP/IP was originally developed as a part of the ARPANET project.

ARQ *see* Automatic Repeat Request.

AS *see* Abstract Syntax.

ASCII (AMERICAN STANDARD CODE FOR INFORMATION INTERCHANGE) A seven-bit coded character set (8 bits when parity check is included) used for representing characters in computer systems, communications systems and related devices.

ASCII FILE TRANSFER PROTOCOL This protocol is used to send or receive files that contain only normal ASCII characters. It sends data one character at a time, provides no error checking and frequently uses XON/XOFF for handshaking. This protocol cannot be used for binary files.

ASE *see* Application Service Element.

ASN.1 Abstract Syntax Notation One. A way of describing protocol standards or protocol functions in the abstract.

ASPECT SCRIPT LANGUAGE A high-level communications programming language used by the PROCOMM communications software package.

ASR *see* Automatic Send and Receive.

ASSOCIATION In the OSI Reference Model, this term is the equivalent of a connection at the application layer. Specifically, the use of a presentation layer connection by the application layer.

ASSOCIATION CONTROL SERVICE ELEMENT *see* ACSE.

ASSOCIATION OF RESEARCH LIBRARIES (ARL) An organization of the major research libraries in the United States and Canada.

ASYMMETRICAL DUPLEX TRANSMISSION A situation in which transmission goes in two directions via a circuit but not concurrently.

ASYNCHRONOUS BALANCED MODE *see* ABM.

ASYNCHRONOUS RESPONSE MODE (ARM) An HDLC mode for logical point-to-point connections.

ASYNCHRONOUS TRANSFER MODE (ATM) A data communications technique used for packet transfer in which units of data, of fixed length, travel through the switch fabric in route to their final destination. This technique is often being used for packet switching on fiber optic networks and is sometimes called cell relay. *see also* Switch Fabric; Frame Relay; Fast Packet Switching.

ASYNCHRONOUS TRANSMISSION A data communications technique which occurs when bytes (characters) of information are sent with unequal time intervals between them with special start and stop bits which are identifiable by the sending and receiving devices. Since characters can be sent one at a time, this transmission technique requires a less sophisticated interface than a synchronous one. Contrast with Synchronous Transmission.

AT COMMAND SET The set of commands originally developed by Hayes, Inc. for the control of modems. The Hayes or "AT" command set have become the industry standard for modems at speeds from 300-9600 baud. The term "AT" is derived from the fact that each of the commands in this protocol begin with the letters AT.

ATC *see* Air Traffic Control.

ATM *see* Asynchronous Transfer Mode.

ATN *see* Aeronautical Telecommunications Network.

ATOMIC ACTION An operation or task which must be fully completed or else not done at all.

ATTENUATION The amount by which an electrical signal (including microwave, radio, infrared, and light waves) weakens over distance as it moves through the transmission media.

ATTRIBUTES Values associated with an object. For example, a file will have attributes such as a name, owner, data type, etc.

AU An abbreviation for "access unit."

AUDIO TELECONFERENCE A conferencing system employing voice-only communications.

AURORA One of five gigabit network research testbeds for NREN. It is on the East Coast of the U.S. and has planned investigations into multimedia systems and distributed shared memory.

AUSEAnet The Australasia and South East Asia Network which was founded in 1986 to support a Very Large Scale Integration (VLSI) project in Thailand, Indonesia, Malaysia, Singapore, Brunei, the Philippines and Australia.

AUTHENTICATION A security technique which proves a user is who he/she claims to be.

AUTHORITY AND FORMAT IDENTIFIER (INDICATOR) (AFI) The first field element of an NSAP (Network Service Access Point) address which indicates the addressing domain and format of the rest of the address.

AUTO-ANSWER The ability of a modem to automatically detect an incoming phone call and provide the appropriate carrier signal.

AUTO BAUD DETECT The ability for a communications device (e.g. modem) and appropriate software to detect and automatically change its communications speed (baud rate) in response to an incoming call.

AUTO-DIAL The ability of a computer to automatically dial a telephone number through a modem.

AUTOMATIC ALTERNATIVE ROUTING A technique used to reroute telephone calls or data transfer to a different communications path when the primary route is blocked or busy.

AUTOMATIC CALL DIRECTOR (ACD) A unit which directs incoming phone calls to the appropriate location within an organization. This is typically done by requiring the caller to indicate the final destination by pressing a particular key on a touch-tone (dual-tone multifrequency) phone.

AUTOMATIC GAIN CONTROL (AGC) A device which compensates for variations in signal strength in a radio signal by increasing the gain on an amplifier.

AUTOMATIC NUMBER IDENTIFICATION (ANI) The ability for the telephone company to identify the calling party for billing purposes or to let the called party know who is calling.

AUTOMATIC REPEAT REQUEST (ARQ) An error correction routine in which if blocks of data are found to have errors, a request is returned to the sender asking for retransmission of the data.

AUTOMATIC SEND AND RECEIVE (ASR) The ability of a microcomputer or intelligent terminal to store incoming or outgoing messages.

AUTOMATIC SEQUENTIAL CONNECTION A facility supported in many public and private data networks in which one computer (or terminal) can contact other computers or terminals in a predetermined sequence of addresses.

AUTOMATIC TRANSFER The ability to transfer an incoming phone call or terminal from one exchange to another or from one phone or terminal to another.

AVAILABILITY The percentage of time that a computer system or network is operating as expected.

AVIATION VHF PACKET COMMUNICATIONS (AVPAC) A subnetwork access protocol used for air-to-ground communications on the Aeronautical Telecommunications Network (ATN).

AVPAC *see* Aviation VHF Packet Communications.

AWG An abbreviation for "American Wire Gage."

B-CHANNEL In an ISDN network this is one of the communications channels which operates at 64 Kbps.

BACK SPACE (BS) A function in which a communications device can move the cursor backwards for one position.

BACKBONE CIRCUIT *see* Interlocation Trunking.

BACKBONE NETWORK A high speed network that links smaller or lower-speed networks.

BACKBONE ROUTING The routing of data traffic via the main high speed network.

BACKBONE TRUNK *see* Interlocation Trunking.

BACK-HAULED CALL A phone call which is routed to a remote operations center and sent to its final destination using the long-distance facilities of another organization.

BAGNet The Bay Area Gigabit Network is a commercial service being developed by Pacific Bell in which a T3 ATM telecommunications network is being introduced in the San Francisco Bay Area to provide service and act as a network testbed.

BALANCED LINE A transmission line in which there are at least two conductors in which the sum of the electrical currents cancel each other out. Data are transmitted via current flowing in opposite directions in the separate conductors.

BALUN Balanced/Unbalanced. An impedance matching device used to connect balanced twisted-pair cabling with unbalanced coaxial cable.

BAND PASS FILTER A filter that allows only certain frequencies to pass along a circuit.

BAND SPLITTER A device in a multiplexer which allows signals to be split into several narrower band subchannels within the available bandwidth of the multiplexer.

BANDPASS FILTER A circuit which allows a single band of frequencies to pass through it but any frequencies above or below this window are suppressed.

BANDWIDTH The range of frequencies that can be transmitted in a communications medium (cable, fiber, radio, etc.). The difference between the highest and lowest frequencies that can be transmitted is the bandwidth. The higher the bandwidth, the more data that can be transmitted.

BARRNet The Bay Area Regional Research Network connected to the NSFNET. Founded in 1986, this network connects academic, private and government institutions in the San Francisco area and northern California

BAS *see* Basic Activity Subset.

BASEBAND A single channel network that transmits digital signals (*see* Broadband). The transmission is done using time-division multiplexing in which only a single transmission occurs at any one time. An example is an Ethernet data link channel.

BASEBAND MODEM A modulator/demodulator (modem) which has more limited features than a modem needed for the public switched telephone network. These are often used for shorter distances or on private lines.

BASIC ACTIVITY SUBSET (BAS) In an OSI network this term refers to a combination of session layer functional units used to support various activities such as the transfer of electronic messages.

BASIC COMBINED SUBSET (BCS) In an OSI network this term refers to a combination of session layer functional units used to provide minimal support of connections in half and full duplex connections.

BASIC ENCODING RULES (BER) The set of rules or definitions for encoding data handled by the Presentation Layer as specified for encoding options in various ISO and CCITT protocols (e.g. ISO-8823).

BASIC EXCHANGE TELEPHONE RADIO (BETR) A radio telephone in which users can access the local telephone system via a phone patch. It is often used in remote locations where stringing phone lines would be too expensive.

BASIC RATE INTERFACE (BRI) This term is usually used in conjunction with ISDN (BRI is the same as 2B+D) to designate the interface for connecting terminals, microcomputers, telephones and other devices to the communications network. A BRI typically includes one 16 Kbps D-channel and two 64 Kbps B-channels.

BASIC SERVICE ELEMENT (BSE) The group of network services which must be provided by local telephone companies to enhanced service providers. These offerings, which must be unbundled and provided under the provisions of the FCC Computer Inquiry III, are called basic service elements.

BASKET A collection of services which can be offered for one group price.

BATH INFORMATION AND DATA SERVICES *see* BIDS.

BAUD RATE One signal element transmission per second. A baud and Bps are not necessarily the same although they may be. With various modulation techniques it is possible to encode more than one bit of data into a single signal element. Compare with Bps.

BAUDOT CODE A code for the representation of data in which one character is five equal-length bits. This coding scheme has frequently been used by teletype machines or TDD (telecommunications device for the deaf) devices for the hearing impaired.

BBS *see* Bulletin Board Service.

BCNET The British Columbia Regional Network was founded in 1987 to support data communications networking for education, research and technology transfer. It is a member of the CAnet backbone in Canada.

BCC *see* Block Check Character.

B-CHANNEL A 64 Kbps telecommunications channel which may be used for voice, circuit or packet-switched data transfer. It is the basic component of ISDN.

BCS *see* Basic Combined Subset.

BEL A special control character which usually rings a bell or other noise making sound on a terminal or I/O device.

BELL 103 The protocol originally developed by AT&T for 300 baud modems in full duplex over the general switched telephone network.

BELL 201B The AT&T communications protocol which supports synchronous data transmission, full duplex operation over 4-wire leased lines and half-duplex operation over 2-wire leased lines at speeds up to 2400 bps.

BELL 201C The AT&T communications protocol which supports synchronous data transmission, half duplex operation over 2-wire dial-up lines at speeds up to 2400 bps.

BELL 202 The protocol originally developed by AT&T for 1200 baud modems in half duplex over the general switched telephone network.

BELL 208A The AT&T communications protocol which supports synchronous data transmission, full duplex operation over 4-wire leased lines, or half duplex operation over 2-wire leased lines at speeds up to 4800 bps.

BELL 208B The AT&T communications protocol which supports synchronous data transmission, full duplex operation over 2-wire dial-up lines at speeds up to 4800 baud.

BELL 212A the protocol originally developed by AT&T for 1200 baud modems in full duplex over the general switched telephone network.

BELL OPERATING COMPANY (BOC) When AT&T was divested in 1982 as a result of antitrust action by the United States Department of Justice, 22 local Bell operating companies were created. Each of these companies is now independently controlled by one of the seven regional Bell holding companies.

BELL SYSTEM Prior to divestiture in 1982, this term referred to AT&T and its 22 BOCs. This term is also sometimes used to refer to the telecommunications infrastructure of the phone system in the United States.

BER *see* Basic Encoding Rules; Bit Error Rate.

BERKELEY SOFTWARE DISTRIBUTION *see* BSD.

BETR *see* Basic Exchange Telephone Radio.

BIBLIOGRAPHIC INFORMATION TECHNOLOGIES *see* BRS.

BID An attempt in a switched telecommunications system to successfully establish a circuit.

BIDIRECTIONAL COMMUNICATION A communications network in which both users can send and receive data at the same time. *see also* Full Duplex.

BIDS Bath Information and Data Services. A system which offers fixed-cost access to a variety of online databases including ISI products (e.g. Science Citation Index, Social Science Citation Index, Humanities Citation Index, ISTP), Embase (Excerpta Medica) and other files. Via networking over JANET, the system is available to over 60 U.K. universities.

BIG BLUE A nickname for IBM Corporation.

BINARY DIGIT An information element which refers to one of two possibilities. Often these two digits are referred to as 0 or 1.

BINARY FILE FORMAT A file which is stored and transmitted as single binary digits (zeros and ones). In many cases, files are stored in this format for transfer across a network so that it retains the exact format in which it was originally stored.

BINARY SYNCHRONOUS COMMUNICATIONS (BSC) A half duplex data communications protocol in which control characters and procedures manage data transfer.

BIND The UNIX implementation of the Berkeley Domain Name Server software.

BIONET An alternate USENET category for newsgroups relating to issues of interest to biologists (e.g. bionet.molbio.genome-program)

BIPOLAR TRANSMISSION A technique used in digital circuits to reduce the effect of a direct current (DC) component in the binary signal. Unequal numbers of 0s and 1s are sent which creates an additional DC voltage on the circuit. In bipolar transmission, a 1 is represented by either a positive or negative voltage while a 0 is represented by no voltage. By properly using positive and negative voltage values for the 1, the average voltage value of the signal can be kept at zero.

BIS *see* Boundary Intermediate System.

BISDN *see* Broadband Integrated Services Digital Network.

BISYNC (BINARY SYNCHRONOUS) An IBM communications protocol in which a preset group of control character sequences is used for the synchronized sending of binary coded data.

BIT *see* Binary Digit.

BIT ERROR RATE (BER) The ratio of the number of bits incorrectly received to the total number of bits transmitted.

BIT MAPPED A screen display in which the image on the screen is generated and refreshed using a binary matrix (bit map) at a specific location in memory.

BIT ORIENTED PROTOCOL (BOP) Data link controls which place no restrictions on the information content of transmissions (i.e. HDLC, SDLC, ADCCP).

BIT STUFFING A technique used in time division multiplexing to add bits to a data stream to ensure that one channel coincides with the clock of another.

BIT SYNCHRONIZATION An electronic clock which maintains the timing sequence between bits for two communicating devices in a synchronous circuit.

BITCOM A commercial microcomputer-based communications software package.

BITFTP A service run a Princeton University which supports BITNET users who are not on the Internet to retrieve files from Internet FTP sites.

BITNET Because It's Time Network (or Because It's There Network). An electronic mail system that has been established between academic institutions through the auspices of EDUCOM. Founded in 1981, this network was originally created to allow collaboration among systems and programmers at campus computing centers. Although it is not a part of the Internet, connections allow the transfer of email and other data between BITNET and the Internet. BITNET is now run by the nonprofit Corporation for Research and Educational Networking (CREN) which operates BITNET for its academic members for electronic mail and other file transfers. Academic sites can connect to BITNET with a dedicated line to the nearest geographic BITNET site by running the BITNET RSCS network protocols. Or they may use BITNET II which enables sites to run BITNET protocols on top of Internet routing and network management protocols. *see* CREN/BITNET.

BITNET BSMTP This is an adaptation of the SMTP (simple mail transfer protocol) for the BITNET environment.

BITNET LISTSERV PARTICIPANT LISTS A listing of all participants and their email addresses from a LISTSERV server.

BITNET NAME SERVERS A directory of users for the BITNET network maintained at several BITNET sites around the United States.

BITS PER SECOND *see* BPS.

BIZ An alternate USENET category for newsgroups relating to business products and services (e.g. biz.dec.workstations).

BLACKBOX A general term referring to a piece of communications equipment which performs a set of functions.

BLANCA One of five gigabit network research testbeds for NREN. It is a U.S. coast-to-coast testbed which will provide implementation fabric for atmospheric modeling and visualization and an experimental multimedia digital library.

BLOCK A group of data which are transmitted as one unit. Error detection is often provided for the entire block of characters.

BLOCK CHECK CHARACTER (BCC) A verification algorithm used in the block transmission mode.

BLOCKING The act of denying access to a network, computer system or function within a system. The term is also used to refer to the process of combining two or more NSAPs into one NPDU for OSI networks.

BLUE BOOK The file transfer protocol is defined in the "Blue Book" for users on the Joint Academic Network (JANET).

BOC *see* Bell Operating Company.

BOP *see* Bit Oriented Protocol.

BOUNDARY INTERMEDIATE SYSTEM (BIS) An ISO router used for connecting different administrative domains in a network.

BPS Bit Per Second. Transmission at one bit per second. Compare with baud rate.

BR (Boundary Router) *see* Boundary Intermediate System.

BREAK COMMAND This command interrupts communications with another computer system through the sending of an appropriate control character or escape sequence.

BRI *see* Basic Rate Interface.

BRIDGE A device similar to a gateway except it connects similar networks to one another and is normally self-programmed. Bridges function at the data link layer of the OSI model, specifically the media access control (MAC) sublayer. A major advantage of bridges is that any type of protocols being used on the subnets being linked can be forwarded whether they be TCP/IP packets, OSI packets, or whatever. Some bridges can be programmed to filter the packet traffic based on some characteristic, such as source or destination address, packet type or size.

BROADBAND A multiple channel network that carries information by carrier waves rather than directly as digital signals providing greater capacity. Frequency division multiplexing is often used on a broadband to provide data transmission on several channels. Such backbones can carry data, radio, video and voice signals.

BROADBAND INTEGRATED SERVICES DIGITAL NETWORK (BISDN) The major difference between ISDN and BISDN is that the later offers higher speed and broader bandwidth.

BROADCAST A form of transmission in which all subscribers or users on a particular service receive the same information.

BROADCAST STORM The uncontrolled dumping of data onto a network saturating the circuit.

BROUTER A combination bridge and router in which some of the capabilities of a router are built into a bridge.

BROWSER A line-oriented front-end to the World Wide Web (WWW) as an alternative to graphical client software.

BRS A commercial information utility providing access to a variety of databases as well as the leasing of search software. BRS is owned by InfoPro Technologies (previously named Maxwell Online, Inc.) which is a subsidiary of Macmillan, Inc.

BS *see* Back Space.

BSC *see* BISYNC; Binary Synchronous Communications.

BSD Berkeley Software Distribution. A version of UNIX developed by the University of California at Berkeley.

BSE *see* Basic Service Element.

BSMPT *see* BITNET BSMPT.

BTW An abbreviation for "by the way."

BUFFER A temporary memory storage location in a computer or peripheral device from which data will soon be transmitted or received.

BULLETIN BOARD SERVICE (BBS) A set of publicly available services (e.g. electronic mail, files for transfer, teleconferencing) in which users log-on to a host computer to get or share information.

BUNDLED SERVICES Services which are only available from a provider or vendor as a package and not as separate entities.

BURSTY TRAFFIC Data traffic which occurs in bursts going from low to high volume.

BUS One or more conductors used for transmitting power, data or signals. In LANs this is a topology in which devices are attached to a single strand of cable running between two points.

BUSY TONE The sound heard by the user of a telephone when the number being called is already in use.

BYPASS The sending of long distance messages over nonpublic lines, although the transmission could have been done over public facilities.

BYTE Most commonly a string of 8 bits which represent one character.

BYTE SERIAL TRANSMISSION A method of sending data in which bytes are sent in a successive predetermined sequence.

CABLE One or more conductors within a protective sheath. It is used to allow the use of conductors separately or in groups.

CABLE NETWORK A network for the transmission of television or radio broadcast signals over coaxial cable. Several of the major cable network companies are in the processing of upgrading some or all of their users to a fiber optic based network.

CACHE In local area networks this refers to the amount of RAM (Random Access Memory) set aside to hold data that may be accessed again.

CALL The general term used to describe the process of communication between two or more users.

CALL ACCEPTED PACKET In a packet switching network, this is the response given by a terminal to indicate it is ready to receive a call request from another device.

CALL BACK The ability for a communications device (e.g. terminal or telephone) to try to reconnect to the other party after a busy signal has been received.

CALL BLOCKING If all communication channels in a circuit are in use then any new call will be blocked.

CALL CLEARING The procedures and activities required associated with the orderly disengagement of two terminals or other communication devices.

CALL ESTABLISHMENT The activities and procedures associated with the orderly connection of two terminals or other communication devices.

CALL FORWARDING A service in which a telephone call can be sent or forwarded to a number other than the one that was dialed.

CALL PROGRESS SIGNAL A signal between data terminal equipment (DTE) and data communications equipment (DCE) to notify the DTE of the progress in making a call.

CALL STORE FUNCTION A feature often provided in central office switches for PBXs which record and keep information about calls received or placed. This may include the telephone number of the originating station, the time and duration of a call, the called station number and so on.

CALL WAITING An indication by a light on a telephone that calls are queuing waiting to be answered. Common carriers also provide a service in which a call directed to a busy station is held while a special tone is sent to the busy station.

CALREN California Research and Education Network. A project by Pacific Bell (PacBell) to build a fiber-based "gigabit" network to link California universities, research labs, major hospitals, and leading high-tech firms in the San Francisco Bay Area and Los Angeles. The project will not only serve as a testbed for network applications development, but serve as a functioning tool for teleconferencing, medical applications, libraries and university research. It will make use of new digital technology such as ATM (asynchronous transfer mode) which can simultaneously route voice, data and video over a gigabit network.

CAnet Canada Network. This network was founded in 1989 and provides a high-speed backbone linking mid-level networks in Canada. For a while the network was known as NRCnet after the National Research Council, the funding agency. This network plays a similar role in Canada as the NSFNET in the United States.

CANONICAL NAME A mnemonic address used to easily identify a machine address on a network. For example the numeric address for Merit on the Internet is 35.1.1.42 while the canonical name would be merit.edu

CARINET A communications network for business and development organizations with primary nodes in Latin America and the Caribbean.

CARL *see* Colorado Alliance of Research Libraries.

CARL Systems Inc. The developer and marketer of the CARL integrated library system software which operates on Tandem Computers. This for-profit company was spun-off from the nonprofit Colorado Alliance of Research Libraries in 1988. *see also* Colorado Alliance of Research Libraries; UnCover.

CARRIER A signal on a communications medium that does not carry information until it is modified in some manner (e.g. amplitude modulation or frequency modulation). The term is also generically used to refer to a company providing communications services.

CARRIER COMMON LINE CHARGE (CCLC) *see* Access Charge.

CARRIER FREQUENCY The center or average frequency of a carrier signal.

CARRIER SENSE MULTIPLE ACCESS WITH COLLISION DETECTION *see* CSMA/CD.

CARRIER SIGNAL A signal on a communications medium that does not carry data until it is modified in some way (frequency modulation, amplitude modulation).

CASA One of five gigabit network research testbeds for NREN. CASA will have end-points in California and New Mexico to extend parallel computing models using the network to carry out large scale scientific and engineering computations.

CASE Common Application Service Elements. Part of the OSI application layer 7.

CATALIST A software program developed for MS-DOS computers running Windows 3.x which provides users a hypertext version of the OPAC (online public access catalog) directory of the University of North Texas's *Accessing On-Line Bibliographic Databases* produced by Billy Barron.

CATIENET Centro Agronomico Tropical de Investigacion y Enseñanza (CATIE). The Tropical Agricultural Research and Training Center provides the exchange of agricultural and forestry information in Costa Rica, Dominican Republic, El Salvador, Guatemala, Honduras, Nicaragua, and Panama.

CATV Cable Television. The term originally stood for community antenna television.

CAUSE The Association for Managing and Using Information Technology in Higher Education. This organization has long been interested in networking and telecommunications issues in higher education. Although the Association still uses CAUSE as its official name (which originally stood for the College and University Systems Exchange), it no longer is considered an acronym.

CBIS A CD-ROM LAN networking product produced by CBIS, Inc. (Norcross, GA). It is NETBIOS compatible and uses Network-OS as its operating system.

CCA An abbreviation for "conceptual communications area."

CCITT Comite Consultatif Internationale de Telegraphique et Telephonique. An international body established under the United Nations within the International Telecommunications Union (ITU), which develops standards. CCITT standards are prefaced by the letter "V" in their numbering scheme.

CCL *see* Common Command Language.

CCLC *see* Carrier Common Line Charge.

CCR An abbreviation for "Commitment, Concurrency and Recovery."

CCRE *see* Commitment, Concurrency, and Recovery Element.

CCS *see* Common Channel Signaling.

CD An abbreviation for "collision detection." It also is a common abbreviation for "compact disc."

CD-NET A CD-ROM networking product produced by Meridian Data, Inc. which is based on the Novell local area network architecture.

CDDI Copper Distributed Data Interface. A technology developed by Crescendo Communications Inc. which provides data communications at speeds up to 100 Mbps over standard copper twisted-pair wiring. The CDDI technology is meant to be interfaced with a fiber optic network (FDDI) backbone. *see also* FDDI.

CDM An abbreviation for "code division multiplexing."

CDNnet Canadian Research and Education Network. Founded in 1986, this network provides network access for research, educational and advanced development agencies in Canada. It was the first network to actively use the CCITT X.400 messaging protocols.

CD-ROM Compact disc read only memory. An optical storage technology on which data, audio or video can be stored. A standard 5 1/4 inch disk can store up to 600 megabytes on one side.

CELL A geographic area served by a low-powered transmitter for cellular phone technology. In asynchronous transfer mode (ATM) technology, this term refers to the fixed length data element which travels through the switch fabric.

CELL RELAY *see* Asynchronous Transfer Mode.

CELLULAR RADIO A radio communications topology in which geographic areas are divided into smaller areas (cells) which can be served by a centrally located antenna and repeater for the forwarding of radio telephone calls. Mobile radio telephones passing from one cell to another are automatically linked to the repeater and antenna within their region for connection to the public phone system.

CENTRAL OFFICE The location where common carriers terminate customer lines and locate switching equipment to interconnect those lines with other networks.

CENTRAL OFFICE SWITCH The placement of switching equipment in a centralized location.

CENTRAL PROCESSING UNIT (CPU) The main computer or computer chip in a network or computer.

CENTRAL SUPPORT UNIT A department that has the staff and technical expertise to provide high-level support for a given computer or communications system.

CENTRALIZED In a telecommunications network this refers to a network in which all transmissions are to and from a central host or node.

CENTREX A telephone company service offering the features of the telephone central office similar to those provided by a PBX.

CEPT An abbreviation for "Conference of European Postal and Telecommunications Administrations."

CERFnet The California Education & Research Federation Network connected to the NSFNET. This mid-level network serves southern California and was founded in 1988.

CERN The Network of the European Laboratory for Particle Physics. This worldwide physics network is based in Switzerland. This organization developed the World Wide Web software which it has now put in the public domain.

CGM An abbreviation for "computer graphics metafile."

CHAIN A series of communications circuits which are connected together for some purpose.

CHANGEBACK The process of restoring data traffic back to a regular circuit after it has been rerouted for some reason.

CHANGEOVER The process of rerouting data traffic to a different communications circuit because an existing circuit has problems.

CHANNEL In data communications, a one-way path along which signals can be sent between two or more points. In telecommunications, a transmission path (may be one or two way) between two or more points provided by a common carrier.

CHANNEL ATTACHED DEVICE A peripheral unit connected to the input/output channel of a central computer.

CHANNEL CAPACITY The volume of data that can be transmitted over a particular circuit or line.

CHANNEL SERVICE UNIT (CSU) A device which is used with digital data channels to terminate a circuit and that provides testing capabilities, transient power protection, local-loop equalization and other capabilities.

CHANNEL SIGNALING SYSTEM NO. 7 (CSS7) A CCITT standard for common interchange signaling for ISDN.

CHARACTER FRAMING A data transmission method in which data are synchronized between the sending and receiving station for only the duration of sending a single character and not between characters. Each character is "framed" by a start and stop bit.

CHARACTER MODE TERMINAL A terminal operating in an asynchronous mode which uses a technique of framing each character with a start and a stop bit.

CHARACTER SET A collection of characters and the code formats by which they are represented in an electronic system (e.g. ASCII, EBCDIC).

CHAT MODE The ability to carry on a real-time discussion via the keyboard with another user in an online system. Many online systems and bulletin boards have a chat feature.

CHEAPERNET A name variation for Thin Ethernet.

CHECK BIT A bit associated with a block of data which is used for checking data transmission errors in the character or block.

CHECK DIGIT An arithmetic process is applied to a pattern of digits and appended to the character string to ensure the proper transmission of data.

CHECKSUM A group of data items are added for an error correction routine.

CICNet The Committee on Institutional Cooperation Network (upper Midwest United States). The network was founded in 1988 and connects institutions of higher education in Illinois, Iowa, Michigan, Minnesota, Ohio and Wisconsin.

CIRCUIT A method of bi-directional communications between two or more points.

CIRCUIT SWITCH A unit which initiates, maintains and ends communications connections. One difference between circuit switches and other telecommunications switches is that the when a connection is established the circuit is maintained on a dedicated basis for the length of the connection.

CIS B *see* CompuServe B.

CIX Commercial Internet Exchange (sometimes called the Commercial Information Exchange). An agreement among network providers on the Internet regarding the accounting of commercial traffic on the Internet. CIX was founded in 1991 as a trade association open to all Internet providers and carriers using TCP/IP with the goal of providing connectivity to all TCP/IP carriers with no restrictions on commercial use of the network.

CLARI An alternate USENET category for newsgroups relating to ClariNet which are groups of articles from commercial news sources (e.g. clari.sports).

CLASS A collection or grouping of objects, services or devices.

CLEAR CHANNEL In digital circuits, if the full bandwidth of the communications channel is available for user data it is said that the channel is clear.

CLEAR TO SEND (CTS) A control signal on a modem which indicates that data terminal equipment (DTE) can initiate the sending of data.

CLEARING The process associated with disconnecting a call on a circuit and making it available for the next user.

CLIENT A workstation or computer attached to a network that can be used to access network resources.

CLIENT-SERVER INTERFACE A program operating on a local microcomputer, workstation, or timesharing computer system which provides an interface to remote information systems (e.g. databases). The end-user is insulated from the remote system database access protocols so that a common user interface is supplied to the human. The client/server software splits up the task of retrieving and displaying information. The client handles the interface and display operations on the local computer while the server focuses on supplying information from a database or service.

CLNP Connectionless-Mode Network Protocol. In OSI networks, ISO 8473 defines this protocol which supports data transfer services between peer network entities. The implication is that the functionality of the protocol is independent from the operation of the subnetworks that transfer the data.

CLNS Connectionless-Mode Network Service. A network service provided by a CLNP (connectionless-mode network protocol).

CLOCK A signal provided by a network, computer system or I/O devices to synchronize the transfer of data between equipment.

CLOSED USER GROUP (CUG) A logical grouping of users in an X.25 service which sets up logical barriers between the designed sets of users.

CLTS An abbreviation for "Connectionless Transport Service."

CLUSTER A term often used to designate the close physical proximity of two or more terminals, I/O devices, telephones, workstations, etc.

CLUSTER CONTROLLER Equipment that concentrates a number of terminals or I/O devices and handles the communication requirements to connect with the host computer.

CMIP Common Management Information Protocol. An OSI standard for network management.

CMIS Common Management Information Services. An OSI standard for network management which is used to allow the SMAE to intercommunicate with protocols in an ISO suite for the exchange of status and management information. CMIS services are provided by CMIP.

CMOT An abbreviation for "CMIS Over TCP/IP."

CNI Coalition for Networked Information. An organization formed in 1990 by the Association for Research Libraries (ARL), CAUSE and EDUCOM to focus on topics such as NREN, intellectual property rights, standards, licensing, economic models and other areas relating to computers in higher education, libraries and information technology.

CNIDR Clearinghouse for Networked Information Discovery and Retrieval (pronounced "snider").

CNMA Communications Network for Manufacturing Application. A collaborative European research effort.

COAX Coaxial cable has one or more central wire conductors surrounded by a dielectric insulator and encased in either a wire mesh or extruded metal sheathing. It is a type of electrical cable which has high bandwidth, low cost, and low susceptibility to interference. However, it is often difficult and expensive to install.

CODEC An abbreviation for "coder-decoder." A device that converts an analog signal into digital form for transmission and reconverts it at the signal's destination.

COLLISION A condition which occurs when two or more stations try to transmit data at the same time on a network.

COLLISION WINDOW A period of time in a network in which there can be undetected contention for the physical communications channel.

COLLOCATE The ability of enhanced service providers to place networking equipment in the same central office location of some other company, for example, the telephone company.

COLORADO ALLIANCE OF RESEARCH LIBRARIES (CARL) An association of academic and public research libraries in Colorado and Wyoming. Through this Alliance, the CARL integrated library system was developed which is now being marketed through a for-profit arm, CARL Systems Inc. *see* CARL Systems Inc.

COLOURED BOOK The set of communications protocols developed by the Joint Academic Network (JANET) which are widely used in Britain, Ireland, Australia and New Zealand. Although based on the OSI Reference Model, many of the protocols predate OSI. JANET has a commitment to migrate to OSI protocols.

COM Original high-level domain for commercial organizations (e.g. businesses) in the domain name system on the Internet (e.g. yoyodyne.com).

COMMERCIAL INTERNET EXCHANGE *see* CIX.

COMMITMENT, CONCURRENCY, AND RECOVERY ELEMENT (CCRE) An ISO application layer protocol which provides recovery and renegotiation services to other application layer protocols.

COMMON APPLICATION SERVICE ELEMENTS *see* CASE.

COMMON CARRIER A government regulated private company that furnishes the public with telecommunications services (e.g., phone companies).

COMMON CARRIER BUREAU An office within the FCC which is responsible for regulating telecommunications within the United States.

COMMON CHANNEL SIGNALING (CCS) The use of a pre-agreed upon communications channel for call signaling which is different from the channel that carries the call itself.

COMMON COMMAND LANGUAGE (CCL) The ANSI/NISO Z39.58 standard for specifying a universal search language for bibliographic information retrieval systems. The standard specifies a command driven system which will be useful in providing a uniform means of access across many different online library catalogs and other online systems. Some feel that the standard will have only limited impact since graphical user interfaces and other techniques will make this type of common search interface unnecessary.

COMMON MANAGEMENT INFORMATION PROTOCOL *see* CMIP.

COMMON MANAGEMENT INFORMATION SERVICES *see* CMIS.

COMMUNICATIONS The transmission of information between two or more points, ideally without alteration in sequence or structure.

COMMUNICATIONS ACT OF 1934 A law passed by Congress in 1934 which made the development of a reasonably priced, publicly available and high quality telephone system a national priority. The Act also included the regulation of radio communications and the establishment of the Federal Communications Commission (FCC).

COMMUNICATIONS CONTROLLER A device on a communications network which manages all data link control activities between a host computer and a distributed user pool.

COMMUNICATIONS LINE Any medium for the transmission of information from one point to another.

COMMUNICATIONS NETWORK FOR MANUFACTURING APPLICATION *see* CNMA.

COMMUNICATIONS PROCESSOR A computer which interfaces a data communications network to a host computer or several hosts.

COMMUNICATIONS PROTOCOL A formal set of standards or conventions to govern the transfer of data over a communications media.

COMMUNICATIONS SERVER A stand-alone device which directs and organizes communications from a network to a mainframe and vice versa.

COMP A newsgroup category distributed to USENET sites which covers topics relating to computers (e.g. comp.lang.fortran).

COMPLEX CABLE The required wiring or cables (including the associated equipment and connectors) to support telecommunications for a variety of applications or users such as terminals, telephones, etc.

COMPOSITE LINK A communications circuit carrying multiplexed data. It is also used to mean the circuit or line connecting at least two concentrators or multiplexors.

COMPOSITION CODING A multi-bit coding technique used to support special characters for such things as diacritics and vernacular language displays. This method requires that the character and the associated diacritic be encoded as separate bytes which are composed at the display device.

COMPUSERVE A popular information utility which supports database access, gateways, software downloading, electronic mail (including the ability to send electronic mail to and from Internet users) and other applications.

COMPUSERVE B (CIS B) This file transfer protocol was designed to be used with the CompuServe online information system.

COMPUTER CONFERENCE The ability for several parties to communicate over a computer network on a topic of common interest. Many public computer conferences occur over BITNET and the Internet and are commonly called "lists." The software packages used to support these lists are often referred to as the "list server" or ListServ. Many computer conferences on BITNET use list server software developed by Eric Thomas.

COMPUTER INQUIRY I (CI-I) Between 1966 to 1971 the FCC conducted an examination in which it was concluded that common carriers could provide computer services to unaffiliated entities through separate affiliates.

COMPUTER INQUIRY II (CI-II) Between 1976 through 1982 the FCC studied common carrier activity and concluded that telecommunications and data processing should not be separately defined. The new ruling became effective on January 1, 1982 and established basic and enhanced services.

COMPUTER INQUIRY III (CI-III) In 1986 the FCC began a third inquiry with the basic result that nonstructural safeguards could be exchanged for extant computer inquiry II subsidiary rules.

CONCATENATION The linking together of several data elements into one larger element. For example, the linking of two or more NPDUs into one (N-1) SDU.

CONCENTRATOR A unit which connects a number of circuits that are not all concurrently used to a smaller number of circuits. Also called an asynchronous time division multiplexer.

CONCERT Communications for N.C. Education, Research, and Technology. A private data communications networked owned and operated by the Center for Communications at Microelectronics Center of North Carolina (MCNC). It serves research institutions, government labs, non-profit organizations and commercial concerns in North Carolina.

CONCRETE SYNTAX The real or actual representation of data sent across the network.

CONDITIONING OF A LINE When data are sent over any communications medium, analog waveforms are distorted due to a variety of environmental factors and the medium itself. Conditioning is a technique whereby channel distortion is counteracted by the use of electrical techniques to modify the existing signal to reproduce a signal closer to its original form.

CONFIGURATION CONTROL The functions necessary to properly direct, manage and control data flow through a network.

CONFIRMED A handshake which verifies that a network connection has been made, a data packet received or any approval of a network activity.

CONFORMANCE TESTING SERVICE (CTS) A European Community program for OSI conformance testing.

CONNECT TIME Usually this term refers to the amount of time that a terminal or workstation has been logged on to a computer for a particular session.

CONNECTION The actual linking of two or more entities across a network to provide an agreed upon service.

CONNECTION-ORIENTED MODE NETWORK SERVICE (CONS) A network service sustained by a connection-oriented network protocol such as X.25 or ISO 8208.

CONNECTIONLESS Communications without previous establishment of a connection.

CONNECTIONLESS-MODE NETWORK PROTOCOL *see* CLNP.

CONNECTIONLESS-MODE NETWORK SERVICE *see* CLNS.

CONNECTIVITY The ability of computer devices to send and receive data with one another. However, this does not mean that two devices necessarily inter-operate.

CONS *see* Connection-Oriented Mode Network Service.

CONSER Cooperative Online Serials. A program in which 19 U.S. libraries catalog and authenticate serial cataloging records. The total number of records which have been created in the program now exceed 600,000 titles and they reside in the OCLC Online Union Catalog and they are available on tape for loading on other systems.

CONSTRAINT SET In file transfer and access management (FTAM), this term refers to a limited file operation.

CONTENTION A situation in which multiple users are trying to send data at the same time. The term is also used to indicate several users attempting to make a bid for the same communications channel at the same time.

CONTINUOUS PRESENCE The display of two or more video images at the same time. Images may appear either on one monitor, using a split screen format, or displayed on two different monitors.

CONTROL CHARACTER A special key on a keyboard which when held down simultaneously with another character, generates commands (or control) specific to a particular application.

CONTROLLER A device in a communications network which manages the flow of data by receiving, interpreting and transmitting signals.

COPPER DISTRIBUTED DATA INTERFACE *see* CDDI.

CORPORATION FOR OPEN SYSTEMS (COS) A group of computer companies which are developing support for OSI-based products with a special emphasis on conformance testing.

COS *see* Corporation for Open Systems.

CPE *see* Customer Premises Equipment.

CPU *see* Central Processing Unit.

CR *see* Carriage Return; Connection Request; Call Request.

CRC *see* Cyclic Redundancy Check.

CREN/BITNET Corporation for Research and Educational Networking. CREN is the nonprofit corporation which operates BITNET for its academic members for electronic mail and other file transfers. Academic sites can connect to BITNET with a dedicated line to the nearest geographic BITNET site by running the BITNET RSCS network protocols. Or they may use BITNET II which enables sites to run BITNET protocols on top of Internet routing and network management protocols.

CROSS MODULATION Interference caused by two or more carriers in a transmission system electromagnetically interacting through nonlinearities in the system.

CROSS SUBSIDIZATION The use of revenue or profit from one service to help pay for the cost of other services.

CROSSTALK A leakage of a transmitted signal when a signal on one path escapes onto another transmission path causing distortion. This is also the name of a popular commercially available communications program for microcomputers.

CSIRONET The Commonwealth Scientific and Industrial Research Organization Network in Australia.

CSMA/CD Carrier Sense Multiple Access with Collision Detection. A LAN access method in which if data are being sent by one node and it detects a collision with other data, it waits and then retransmits. This method is used in Ethernet.

CSNET The Computer Science Network connected to the NSFNET (merged with CREN/BITNET)

CSPDN An abbreviation for "Circuit-Switched Public Data Network."

CSS7 *see* Channel Signaling System No. 7.

CSU *see* Channel Service Unit.

CSUnet The California State University Network (CSUnet) is a state-funded network which provides network access to all California State University campuses.

CTS *see* Clear to Send; Conformance Testing Service.

CUG *see* Closed User Group.

CURRENT LOOP A method of interconnecting computer equipment in which the binary character 1 (or mark) is represented by current on the communications line and the binary character 0 (or space) is represented by no current on the line. This technique is sometimes used for connecting terminals to a main computer because the transmission lines can extend over farther distances than for serial transmissions (unless a line driver is used). Current loop circuits are considered older technology and less frequently used today.

CURRENT LOOP CONVERTER A device which converts a serial RS-232 communications signal into a current loop signal or vice versa.

CUSTOMER PREMISES EQUIPMENT (CPE) At one time the telephone company required that all telecommunications equipment must be provided by itself at a customers' site. Customers now have a choice of acquiring equipment from AT&T, RBOCs and other suppliers.

CYCLIC REDUNDANCY CHECK (CRC) An error correction technique whereby the check character is produced by dividing all the serialized bits in a block by a predetermined polynomial and taking the remainder.

DA An abbreviation for "Destination Address."

DAA *see* Data Access Arrangement.

DAP An abbreviation for "Directory Access Protocol."

DARK FIBER Fiber optic cable which has been deployed but is not yet in use.

DARPA Defense Advanced Research Projects Agency. A U.S. federal agency which was responsible for the development of the ARPANET network and the development of the TCP/IP protocols.

DAS An abbreviation for "Dual Attached Station."

DATA ACCESS ARRANGEMENT (DAA) Data communications equipment approved or furnished by a common carrier that permits the attachment of a computer to the common carrier network (usually phone line). These devices are now often integrated into equipment.

DATA BITS A byte typically requires 8 bits to accommodate the 256 characters in IBM's extended ASCII character set. Mainframes and many other types of computer sometimes only require 7 bits to represent a byte because they only use the first 128 characters of the IBM extended ASCII character set. In communications software it is often necessary to indicate the number "data bits" (7 or 8) to represent a byte.

DATA CHANNEL A transmission channel which is used to send data. It may be analog or digital in nature.

DATA CIRCUIT A two-way communications channel.

DATA CLOCK The signal timing element which is necessary for the detection and decoding of a received signal. This can be derived from the incoming signals.

DATA COMMUNICATION NETWORK ARCHITECTURE *see* DCNA.

DATA COMMUNICATIONS EQUIPMENT (DCE) The hardware required to establish, maintain and end a communications session. For example a modem would fall in this category.

DATA COMPRESSION Many algorithms and methods are available to compress data files to make them smaller. This is useful in telecommunications systems since a reduced file size lessens the amount of time and network resources to transmit and receive a file.

DATA ELEMENT A field of information which is part of a larger message.

DATA ENCAPSULATION *see* Encapsulation.

DATA ENCRYPTION STANDARD (DES) An algorithm for data encryption and decryption.

DATA LINK A communications connection between stations engaged in transferring data in which no store-and-forward element is interposed between them. All transmissions sent by one station are received directly by the other.

DATA LINK CONTROL (DLC) IBM's SNA protocol layer which handles error detection, data transmission, and error recovery.

DATA LINK ESCAPE (DLE) A transmission control code that alters the meaning of other characters that follow after it in a transmission sequence.

DATA LINK LAYER In the OSI model, this is layer 2. It accomplishes the error-free exchange of data between two directly connected systems. It is primarily concerned with the detection and correction of errors which arise during transmission.

DATA LINK SERVICE ACCESS POINT (DLSAP) A type of address used to identify the user of the Data Link Service.

DATA NETWORK A set of physical connections that support the transfer of data between end points. The media supporting the connections can vary greatly.

DATA NETWORK IDENTIFICATION CODE (DNIC) The network code and country code fields for an X.121 address.

DATA PBX A private branch exchange (PBX) which provides switching services for data. Often these are used for point-to-point connections for the interconnection of terminals, workstations or host computers within an organization.

DATA RESOURCES INC. (DRI) An information utility which specializes in financial and statistical data.

DATA SERVICE UNIT *see* DSU/CSU.

DATA SET READY (DSR) On modems this is the control signal which indicates to the attached terminal or other peripheral device that the modem is properly connected to the telephone circuit.

DATA SIGNALING RATE The speed at which data are sent over a circuit.

DATA TERMINAL EQUIPMENT (DTE) The peripheral device which acts as a data producer or receiver such as a microcomputer, terminal or telephone headset.

DATA TRANSFER RATE The speed at which data are sent over a communications circuit.

DATABASE VENDOR An organization which provides shared access to a wide variety of databases on a timesharing basis.

DATAGRAM Data which are transmitted as an isolated entity across a network without any prior warning. Conceptually this may be compared to a telegram in that it is a self-contained message that can arrive without notice. This data does not need to be transmitted in order, and in this type of service any error checking must be accomplished by higher-layer protocols or by the end user.

DATAPAC A Canadian packet-switching network service.

DATAPHONE DIGITAL SERVICE (DDS) A communications system in which data are sent in digital rather than analog form. In such a system, there is no need for a modem.

DATATIMES CORPORATION A bibliographic utility specializing in full text newspapers.

DATEX A collection of public data transmission services supported by the Deutsche Bundespost to subscribers in Germany.

DC An abbreviation for "direct current."

DC1 An abbreviation for "device control 1" (XON).

DC3 An abbreviation for "device control 3" (XOFF).

DCC An abbreviation for "data country code."

DCE *see* Data Communications Equipment.

D-CHANNEL The out-of-band signaling channel for ISDN in which control information can be sent to a user outside of the main channels used for information. This control information includes such things as establishing calls between users, changing parameters in equipment or redirecting calls.

DCNA Data Communication Network Architecture. A communications network developed by IBM/Japan and Nippon Telegraph and Telephone Corporation (NTT).

DCS *see* Defined Context Set.

DDD *see* Direct Distance Dialing.

DDN *see* Defense Data Network.

DDS *see* Dataphone Digital Service; Digital Data Service.

DE FACTO STANDARD An unofficial standard that is widely used in the industry because manufacturers choose to use it.

DE JURE STANDARD An official standard developed and adopted by a formal standards-making body such as IEEE, ANSI, CCITT, ISO, etc.

DECMAIL A proprietary electronic mail system developed by Digital Equipment Corporation (DEC).

DECNET Digital Equipment Corporation's proprietary networking protocol to enable the networking of DEC computers. Phase V is OSI compliant.

DECODER A device which takes information in a defined code and generates output in a form required by another processing system.

DEDICATED LINE A dedicated or leased line is a communication line used exclusively by one customer. In a telephone leased-line, the line often does not pass through inter-exchange switching equipment and the line is leased on a flat monthly fee regardless of how much data are transmitted. Also called a private line or a leased line.

DEFENSE ADVANCED RESEARCH PROJECTS AGENCY *see* DARPA.

DEFENSE DATA (DIGITAL) NETWORK (DDN) A portion of the Internet which connects to the United States military bases and contractors. MILNET is one of the DDN networks. This is used for non-secure communications. DDN also runs its own network information center (NIC) where much of the DDN traffic is archived.

DEFENSE TECHNICAL INFORMATION CENTER - *see* DTIC.

DEFINED CONTEXT SET (DCS) In the OSI protocol suite this term refers to a set of defined presentation layer contexts.

DELAY The time required for a data unit to cross a network.

DELAY DISTORTION A type of signal distortion caused by different propagation times for various frequencies in a circuit.

DELAY EQUALIZER A device which compensates for delay distortion in a circuit. *see also* Delay Distortion.

DELPHI A popular information utility.

DEMARC The point of connection of telecommunications wiring within a building to a public carrier. The demarcation point between carrier equipment and customer premises equipment is often at terminal block in a telecommunications room.

DEMODULATION The process of extracting a message signal from the carrier signal in a circuit.

DEMULTIPLEXING The separation of several data streams or common frequency bands on one multiplexed line into individual channels.

DEMULTIPLEXER A piece of equipment which does demultiplexing.

DEMUX An abbreviation for "demultiplexer."

DENet The Danish Ethernet Network provides network access to computers at the Danish Computing Centre and those under the control of the Danish Ministry of Education. It is part of NORDUnet.

DEPACKETIZING The process of stripping the codes which have been added to data for sending data over a packet switching network.

DES *see* Data Encryption Standard.

DESPOTIC NETWORK A network in which the timing of all network traffic is governed by one master clock.

DEVICE CONTROL CODES Special codes within standard character sets which are used to control special functions on a terminal, printer or other I/O device (e.g. DC1, DC3).

DFN Deutsches Foerschungsnetz. German research network founded around 1984 to connect academic and research network throughout Germany. Although the network originally served West Germany, the unification of Germany is bringing connections throughout the entire country. This network is also known as WIN or Wissenschafnetz (Science Network).

DFS For all practical purposes this is another name for AFS (Andrew File System) which offers the ability to share files over the Internet. To be specific, DFS refers to the AFS implementation that is a part of the Open System Foundation's (OFS) Distributed Computing Environment (DCE). *see* AFS.

DIAL UP CONNECTION A connection made to a computer network via a switching operation in an exchange such as the public telephone system.

DIALING DIRECTORY Communications software packages often offer the ability to create a dialing directory to easily connect to remote information systems. This feature typically includes the ability to record the name of the system, phone number, communications parameters, and access to a SCRIPT file for automating certain features (e.g. login).

DIALNET Dialog Information Services' proprietary packet switching network. In the fall of 1993 Dialog turned the management of its U.S. data networking operations over to BT North America's GNS Network (formerly TYMNET).

DIALOG INFORMATION SERVICES A commercial timesharing information utility which offers a wide variety of databases and services. It is a subsidiary of Knight Ridder.

DIALOGLINK A microcomputer-based communications software package optimized for use on DIALOG Information Services.

DIAL-UP LINE A communications circuit which can be accessed from the public telephone network.

DIB An abbreviation for "Directory Information Base."

DIBIT A group of two bits with four possible states which is sometimes used with modems which encode more than one bit for each signal.

DID *see* Direct Inward Dialing.

DIFFERENTIAL PHASE SHIFT KEYING (DPSK) A modulation technique developed by AT&T for its Bell 201 modem protocol in which the phase of the carrier signal is quickly alternated.

DIGITAL CIRCUIT A communications channel which is specifically be designed to handle digital and not analog data.

DIGITAL DATA SERVICE (DDS) A dedicated, 4-wire leased-line circuit capable of transmitting data up to 56 Kbps. DDS lines are linked within the carrier by special repeaters that are maintained separately from analog lines. A channel service unit/data service unit (CSU/DSU) replaces traditional modems as the customer links to this service.

DIGITAL NETWORK ARCHITECTURE *see* DNA.

DIGITAL REPEATER An amplification device which regenerates a digital signal at regular intervals along a communication path to overcome signal attenuation.

DIGITAL SIGNALS Signals which consist of a series of discrete elements that have only one value at a time. Digital transmission systems are the heart of most modern communication systems. Contrast with Analog Signals.

DIGITAL SWITCH A unit which makes switched connections for digital data transmission between different circuits to establish communication paths between systems.

DIGITAL TERMINATION SYSTEM (DTS) A microwave packet-radio technique in which messages can be broadcast from a central antenna to receivers on the roofs of buildings or at other locations.

DIGITAL VIDEO A video signal which has been digitized and is comparable in quality to that of broadcast television, but less expensive to transmit.

DIRECT DISTANCE DIALING (DDD) The ability to make long distance phone calls without using an operator.

DIRECT INWARD DIALING (DID) The ability of a PBX to receive called station identification information from a central point or office over a common trunk facility without going through an operator.

DIRECT OUTWARD DIALING (DOD) The ability of a PBX or other remote device to make calls on the public phone network without going through an operator.

DIRECTORY In networks this term is used to signify a service that offers name-to-address mappings and related functions.

DIRECTORY SERVICE A CCITT and ISO application layer protocol for providing directory services for remote or local users.

DIRECTORY SERVICE AGENT (DSA) A directory service for accessing a remote or local directory information database.

DIRECTORY USER AGENT (DUA) A local agent in the "using application protocol" that accesses the directory service through the DSA (directory service agent).

DIS *see* Draft International Standard.

DISCONNECT SIGNAL A signal transmitted from one end of a data communications line designating that a session should be terminated.

DISK CHANNEL The sole route for data traveling to the hard disk for recording or for data being read from disk and going to memory.

DISKLESS WORKSTATION A microcomputer in a network that lacks floppy or hard disk drives. A diskless device boots the networking operating system from local read only memory.

DISTINGUISHED NAME The global name of an entity.

DISTORTION An undesired alteration of a waveform which occurs during transmission.

DISTRIBUTED RELATIONAL DATABASE ACCESS (DRDA) A standard being developed by IBM to support database access across all of its platforms including those using the SQL data access language.

DISTRIBUTED PROCESSING The interconnection of geographically dispersed computer systems each performing various functions for a common purpose. Control does not reside at one point but is distributed throughout the network.

DISTRIBUTED SERVER A local area network configuration in which there are many file servers spread throughout the workstations, as opposed to a central file server.

DISTRIBUTED SWITCHING A telecommunications topology in which switching functions are not done centrally but at local switches.

DISTRIBUTED SYSTEMS ARCHITECTURE (DSA) A proprietary network architecture developed by Honeywell.

DISTRIBUTED SYSTEMS NETWORK (DSN) A proprietary network architecture developed by Hewlett-Packard.

DIT The abbreviation for "Directory Information Tree."

DIVESTITURE Through an antitrust action initiated by the U.S. Department of Justice against AT&T, the Bell system divested itself of the local portions of its 22 Bell operating companies (BOCS). The initial legal action began in 1974 and culminated in the AT&T signing of the January 1982 Modified Final Judgment or Divestiture.

DIVISION OF LIBRARY AUTOMATION *see* DLA.

DIXIE A protocol defined in RFC1249 which allows TCP and UDP-based clients to access an X.500 OSI Directory Service without using the Directory Access Protocol (DAP), an OSI application layer.

DL *see* Data Link.

DLA Division of Library Automation. An office of the University of California which has developed the MELVYL online union catalog and has done pioneering in library networking and Z39.50 links.

DLC *see* Data Link Control.

DLE *see* Data Link Escape.

DLSAP *see* Data Link Service Access Point.

DMCONNECTION A company which provides low-cost access to UUCP mail exchange and news feeds to the general public. It also provides Internet access to the public and is located in Hudson, MA.

DNA Digital Network Architecture. The network architecture of Digital Equipment Corporation, of which DECnet is one implementation. DNA is similar in intention and concept to the OSI suite, but because it was developed prior to OSI did not offer OSI compliance until DECNET Phase V.

Dnet Founded in 1983, Dnet is the German branch of EUnet.

DNIC *see* Data Network Identification Code.

DNS *see* Domain Name Server.

DOD *see* Direct Outward Dialing.

DOD U.S. Department of Defense. The DOD's Advanced Research Projects Agency created ARPAnet which was seminal in the beginnings of the Internet. *see* ARPANET.

DOE/RECON A bibliographic utility operated by the U.S. Department of Energy specializing in energy and environmental information.

DOMAIN In IBM's Systems Network Architecture (SNA), a mainframe-based systems services control point (SSCP) and the physical units, logical units, links, link stations, and all the associated resources that the host has the ability to control.

DOMAIN NAME SERVER (DNS) A system that provides the translation of computer names into numeric TCP/IP addresses and vice-versa. DNS provides the ability to use the Internet without having to remember long lists of numbers. It uses a hierarchical tree-structured naming scheme and supports communications protocols with nameservers on the network which map alphanumeric names to TCP/IP numeric addresses.

DOMAIN SPECIFIC PART (DSP) The third, and optional, field in a network service access point (NSAP) address which indicates the remainder of the address necessary to identify the user of the network service.

DOMESTIC SATELLITE CARRIER (DSC) A communications service using satellites operating in the United States for intercity traffic.

DOS GATEWAY To temporarily activate the DOS operating system so that a DOS task can be performed without leaving the software program.

DOW JONES NEWS/RETRIEVAL An information utility operated by Dow Jones & Co. Inc. with bibliographic and financial information.

DOWNLINK The portion of a satellite circuit extending from the satellite to an earth station.

DOWNLOADING The process of capturing data being received from another computer in memory or on disk. The term is also used to mean the process of loading software into the nodes of a network from one node over the network medium.

DP *see* Draft Proposal.

DPSK *see* Differential Phase Shift Keying.

DRA Data Research Associates. A major integrated library system vendor which runs software developed for DEC hardware platforms.

DRAFT INTERNATIONAL STANDARD (DIS) An intermediate stage in the approval of ISO standards before a document is recognized as a full International Standard.

DRAFT PROPOSAL (DP) An early stage in the development and approval of ISO standards in which comments are requested from the community. This stage comes before the document is a "draft international standard." *see also* Draft International Standard.

DRANET A proprietary computer network run by DRA which links library systems into a central DRA computer system for database access.

DRDA *see* Distributed Relational Database Access.

DRI *see* Data Resources Inc.

DS *see* Directory Service.

DS0 Digital signal level 0. In telephony this refers to one 64 Kbps standard digital telecommunications channel.

DS1 Digital signal level 1. An AT&T standard for the transmission of high speed data over T1 facilities (1.544 Mbps).

DS-1C Digital signal level 1C. An AT&T standard for the transmission of high speed data over T1C facilities (3.152 Mbps).

DS2 Digital signal level 2. An AT&T standard for the transmission of high speed data over T2 facilities (6.312 Mbps).

DS3 Digital signal level 3. An AT&T standard for the transmission of high speed data over T3 facilities (44 Mbps).

DS4 Digital signal level 4. An AT&T standard for the transmission of high speed data over T4 facilities (273 Mbps).

DSA *see* Distributed Systems Architecture; Directory Service Agent.

DSAP An abbreviation for "Destination Service Access Point."

DSC *see* Domestic Satellite Carrier.

DSIRnet The Department of Scientific and Industrial Research Network which provides computer networking for several government installations in New Zealand.

DSN *see* Distributed Systems Network.

DSP *see* Domain Specific Part.

DSR *see* Data Set Ready.

DSU/CSU The DSU is a synchronous data line driver for short-distances and a CSU is an access arrangement that offers local-loop equalization, protection from electrical transients, self-testing, and circuit isolation. Thus a DSU/CSU is used to connect digital data devices to a digital network. These devices are commonly bundled.

DTE *see* Data Terminal Equipment.

DTE WAITING A control signal between data terminal equipment (DTE) and data communications equipment (DCE), that the DTE device is waiting for another signal from the DCE.

DTIC Defense Technical Information Center. A bibliographic utility operated by the U.S. Department of Defense which specializes in classified documents.

DTMF *see* Dual Tone Multifrequency.

DTS *see* Digital Termination System.

DUA *see* Directory User Agent.

DUAL TONE MULTIFREQUENCY (DTMF) A signaling system used in telecommunications system using the digits 0 through 9 and the special characters # and *. It is also known as touch tone.

DUMB TERMINAL A teletype compatible terminal which operates asynchronously using ASCII code for communicating. These devices generally do not perform storagc functions as is availablc on a microcomputer or intelligent terminal.

DUPLEX CIRCUIT A telecommunications circuit which allows the transmission of data in both directions at the same time. The term "full duplex" is also commonly used.

DYNIX A major integrated library system vendor which is a wholly owned subsidiary of Ameritech. This vendor has an especially strong presence in the medium-sized public and academic library market.

E.164 The network addressing plan for the ISDN as specified by CCITT.

E.166 The recommended addressing plan for the interconnection of different types of public data networks (including ISDN) as specified by CCITT.

EAN The X.400 experimental research network in Spain.

EARN The European Academic Research Network was founded in 1983 as a backbone network connecting national and academic networks in Western European countries as well as Austria, Yugoslavia, Cyprus, Turkey, Israel, Algeria, Ivory Coast, Morocco, Tunisia and Egypt.

EARTH STATION The antenna and associated equipment used for transferring signals to and from a communications satellite.

EBCDIC Extended Binary-Coded Decimal Interchange Code. The 8-bit character code scheme developed and used by IBM.

ECHO A delayed message signal, reflections created by variations in the connection, in a communications circuit which may degrade the quality of the connection.

ECHO SUPPRESSER A device which blocks echoed signals thus eliminating reflected signals in a circuit.

ECHOPLEX An error checking method for data communications in which the received characters are returned (echoed) to the sender for verification.

ECMA European Computer Manufacturers Association.

ED End Delimiter. A field used in the ANSI FDDI and the IEEE 802.5 communication protocols.

EDI *see* Electronic Data Interchange.

EDU Original high-level domain for educational institutions in the domain name system on the Internet (e.g. carl.lib.asu.edu).

EDUCOM An organization involved in the development and use of computers, telecommunications and networking in education. It created and operated EDUNET in the late 1970s and early 1980s to allow institutions to share computing resources and applications. In 1985, EDUCOM formed the Networking and Telecommunications Task Force (NTTF).

EFF *see* Electronic Frontier Foundation.

EFT An abbreviation for "electronic funds transfer."

EGP *see* Exterior Gateway Protocol.

EIA *see* Electronic Industries Association.

EIA-530 A communications interface standard using 25-pins designed for nonswitched applications using wideband channels at speeds of 20 Kbps or greater.

EINET Enterprise Integration Network. A nonprofit TCP/IP based network run by Microelectronics and Computer Technology Corporation (MCC) which is architecturally similar to the Internet but carries data traffic for commercial clients and offers better network security.

EIS *see* Executive Information System.

E-JOURNAL A periodical distributed in electronic form.

ELECTRONIC DATA INTERCHANGE (EDI) A series of standardized message formats which support the electronic transmission of business transactions between computers on a network. Although these message formats were primarily developed for specific sectors of the business community (e.g. banking, grocery, financial community), the general concept has been broadened to include areas such as the electronic ordering of publications. As one example, the ANSI X.12 standard, Data Interchange Standards, is in use between libraries and publishers (or jobbers) for the ordering of books and other publications. The EDI formats contain three basic elements: a data dictionary which defines the contents of data elements; a set of message standards which define the format and data content of messages; and a message syntax which specifies the structure of messages needed to exchange the data.

ELECTRONIC FRONTIER FOUNDATION An organization located in Cambridge, MA which was established to encourage the free and open flow of information in electronic networks with the goal of making these resources available to everyone, not just the technical elite.

ELECTRONIC INDUSTRIES ASSOCIATION (EIA) A standards organization in the United States which specializes in electrical and electronic issues including telecommunications.

ELECTRONIC JOURNAL *see* E-Journal.

ELECTRONIC MAIL (EMAIL) The electronic transmission of messages or documents in a computer system or between computers. Quite often this is done using a store and forward technique.

ELECTRONIC SWITCHING SYSTEM (ESS) Electronic devices which perform the switching functions in a telecommunications system.

ELEMENT MANAGEMENT SYSTEM (EMS) A system that manages network elements. This encompasses a host of devices such as modems, multiplexors and DSUs.

ELM A popular electronic mail system which has been written for the UNIX environment.

EM *see* End of Medium.

EMA Enterprise Management Architecture. Digital Equipment Corporation's OSI-based network management strategy. Like UNMA, EMA provides for multivendor network management using interfaces based on OSI standards.

EMAIL *see* Electronic Mail.

EMS *see* Element Management System.

ENCAPSULATION The placing of a packet of data as a unit of data within another packet of a different protocol.

ENCODER A device which converts digital signals in a transmitting station into a form required to translate physically separate signals of synchronization pulses and data into a single serial bit stream.

ENCRYPTION The process of encoding a bit stream before transmission to provide data security.

END DELIMITER *see* ED.

END OF MEDIUM (EM) A control character which designates the end of a data field.

END OF MESSAGE (EOM) A control character which designates the end of a text message.

END OF TRANSMISSION (EOT) A control character which indicates the termination of the transmission of a group of data.

END OF TRANSMISSION BLOCK (ETB) An internationally recognized control code which indicates the end of a block forming part of a message.

END ROUTING DOMAIN A routing domain specified by ISO which does not allow the relay of PDUs (protocol data units) into other domains because of policy or connectivity limitations.

END SYSTEM (ES) In the context of the OSI protocols this term defines an "open system" that can communicate with other OSI end systems via the OSI protocols.

END SYSTEM HELLO (ESH) A signal (protocol data unit) send by end systems (using ISO 9542) to indicate their presence on a subnetwork.

END-TO-END Communication between the source and destination nodes on a network. The transport layer handles end-to-end communications (contrast with Point-to-Point).

ENERGY SCIENCES NETWORK *see* ESnet.

Enet The Spanish branch of EUnet which was founded in 1986. The "E" stands for España.

ENFIA *see* Exchange Network Facilities for Interstate Access.

ENHANCED SERVICE A service which provides value added features beyond whatever is defined as "basic service." In the telephone industry this often means the use of computers to provide special features beyond point-to-point communications.

ENQ *see* Enquiry.

ENQUIRY (ENQ) A control character used to indicate that a response from a remote system is needed.

ENTERPRISE INTEGRATION NETWORK *see* EINET.

ENVELOPE Information recorded to encapsulate a group of data for proper transmission across a network.

EOM *see* End of Message.

EOT *see* End of Transmission.

EPA An abbreviation for "enhanced performance architecture."

EPIC OCLC Online Computer Library Center's bibliographic online reference service and information retrieval system. It supports the NISO Common Command Language (CCL) as well as Z39.50 information retrieval protocol.

EQUAL ACCESS The legal injunction from the 1982 divestiture of AT&T which required the BOCs to provide other long distance carriers with exchange access services of similar price, type and quality as those provided to AT&T by September 1, 1986.

EQUALIZATION The infusion of controlled signals into a communications channel to offset unwanted distortion.

ERD *see* End Routing Domain.

ERLANG A unit of measurement for rating telecommunications traffic and use demand. One erlang is the intensity at which one traffic path would be continuously occupied (e.g. one call per minute, one call per hour). An erlang equals 36 ccs (36 hundred call seconds) which represents full time use of a conventional telecommunications traffic path.

ERNET Education and Research Network. This network connects academic research institutions in India.

ERROR In data communications this refers to an unwanted change in the original contents of a transmission.

ERROR CHECKING A function performed in most data communication systems to ensure that information is accurately transferred from one site to another.

ERROR CORRECTING CODE Algorithms which contain additional signaling elements to allow the detection and correction of errors in transmission.

ERROR DETECTION The identification of lost or corrupted bits of data in a communications system.

ERROR RATE The ratio of the amount of data incorrectly received to the total amount of data transmitted.

ES *see* End System.

ES-IS An abbreviation for "end system-intermediate system."

ESA/IRS *see* European Space Agency/Information Retrieval Service.

E-SERIAL *see* E-Journal.

ESH *see* End System Hello.

ESnet A national data communications network funded and managed by the U.S. Department of Energy Office of Energy Research (DOE/OER). Its purpose is to support energy research through widespread access to supercomputing facilities. The network is installed and operated by the National Energy Supercomputer Center (NERSC), formerly known as the National Magnetic Fusion Energy Computer Center (NMFECC) located at the Lawrence Livermore National Laboratory.

ESS *see* Electronic Switching System.

ETB *see* End of Transmission Block.

ETHERNET A popular local area network technology invented by Xerox. A passive cable is used to interconnect active devices (e.g. workstations) which communicate using a technology called CSMA/CD (carrier sense multiple access/collision detection) in which if data are being sent by one node and it detects a collision with other data, it waits and then retransmits. It is a comprehensive baseband data communications standard that interconnects computers and local area networks. The access control mechanism is CSMA/CD and the baseband transmission speed is rated at 10 Mbps. The standard version of Ethernet is defined by IEEE in 802.3, 10BASE5. A broadband version is defined in 10BASE36; Thin Ethernet using RG-58 coax cable is defined in 10BASE2, and a version using twisted pair cable is specified in 10BASET.

ETHICS *see* Network Ethics.

ETX An abbreviation for "end of text."

EUDORA An electronic mail system for the Macintosh computer developed at the University of Illinois. It is available on the Internet as shareware and is in wide use at academic institutions. A commercial version is also available.

EUnet The network of EUUG, the European Unix User Group. This is a cooperative news and email network used for research and development. It serves sites in Western Europe and originally began as an extension of the USENET and UUCP systems in North America.

EUnet IN ICELAND This network provides leased, dial-up and X.25 communication links throughout Iceland for UUCP UNIX users. It provides a connection to NORDUnet.

EURONET A European X.25 packet switching network that was commissioned by the European Economic Community (ECC).

EUROPEAN SPACE AGENCY/INFORMATION RETRIEVAL SERVICE (ESA/IRS) One of the major bibliographic utilities in Europe based in Rome, Italy.

EVEN PARITY A condition which exists when all the bits in a row or column of a block of digital data add up to an even number.

EWOS An abbreviation for "European Workshop for Open Systems."

EXCHANGE In the telephone system, this term refers to the geographical area that is billed by a telephone company according to a single charge rate, regardless of political boundaries.

EXCHANGE ACCESS SERVICES Interexchange (i.e. long distance) carriers (e.g. telephone companies) have access to a variety of services from the BOCs that allow interLATA communications.

EXCHANGE AREA A geographic area often equivalent to a Standard Metropolitan Statistical Area (SMSA) in which there is a single set of charges for telephone service. This is called a Local Access and Transport Area (LATA).

EXCHANGE INFORMATION *see* XID.

EXCHANGE NETWORK FACILITIES Facilities within an exchange area which use common trunks, hookups and special-service circuits for transmission.

EXCHANGE NETWORK FACILITIES FOR INTERSTATE ACCESS (ENFIA) A fee charged for interstate communications in which other common carriers can use the exchange network facilities of BOCs.

EXCHANGE TERMINATION In the ISDN arena this indicates the central office link for the end user.

EXECUTIVE INFORMATION SYSTEM (EIS) A system which can store, view and analyze information that is stored in a variety of formats (e.g. spreadsheets, databases) to optimize information management for executives. EIS systems are often developed in a client/server environment where data can be extracted from a variety of hosts and presented on a single client workstation.

EXPAND The proprietary networking protocol used by Tandem Computers Inc.

EXPEDITED DATA The process of moving data through a network as quickly as possible without regard for flow control. In the ISO protocols, this term refers to the transfer of an additional small protocol data unit (PDU) outside of the window values used for flow control. This PDU would be related to control functions and not the transfer of applications information.

EXTENDED BINARY-CODED DECIMAL INTERCHANGE CODE *see* EBCDIC.

EXTENDED NETWORK A communications network which consists of many subnetworks connected by gateways, bridges or routers.

EXTERIOR GATEWAY PROTOCOL (EGP) A routing protocol used by the U.S. military for connecting different types of networks.

EXTERNAL MODEM A stand alone modem, as contrasted with an internal modem which is integrated inside of a computer or terminal.

FACILITY A transmission path between two or more points offered by a common carrier; any or all of the physical elements of a plan to provide communications services.

FACSIMILE *see* FAX.

FADU *see* File Access Data Unit.

FAQ Frequently Asked Questions. A list of frequently asked questions and their answers. Many USENET news groups and other lists maintain FAQ files so that new users to the service do not have to keep asking the same questions.

FARNET *see* Federation of American Research Networks.

FAST CIRCUIT SWITCHING A situation in which a user may directly dial a data transmission facility using a data circuit provided by a common carrier without having to use more expensive leased (or dedicated) lines.

FAST PACKET SWITCHING A variation on traditional packet switching techniques that uses high speed transmission lines for wide area networks. CCITT has developed the I.121 standard for fast packet switching. The high speed transfer of packets is partly due to a reduction in the processing time on intermediate links in the network in the switch fabric. *see also* Frame Relay; Asynchronous Transfer Mode.

FAST SELECT The X.25 facility which supports transaction processing.

FAULT TOLERANT The ability of hardware or software to properly continue to operate even if a failure occurs.

FAX An abbreviation commonly used for facsimile transmission (telefacsimile). It represents the technology used to digitally transmit graphic material over the public telephone network.

FC *see* Frame Control.

FCC Federal Communications Commission. A U.S. government agency which regulates many aspects of communications.

FCS An abbreviation for "frame check sequence."

FD *see* Full Duplex.

FDDI Fiber Distributed Data Interface. This is ANSI standard X3T9.5 and was developed to allow a cluster of large computers to communicate with each other and peripheral devices with 100-Mbps bandwidth for up to 500 station connections over an optical fiber. Dual rings can support up to 1000 nodes.

FDM *see* Frequency Division Multiplexing.

FDUX *see* Full Duplex.

FDX *see* Full Duplex.

FEDERAL INTERNET EXCHANGE (FIX) An organization whose purpose is to improve communications, provide interconnection between different Federal networks and to provide operational and technical assistance to its members (who are U.S. federal agencies) on issues relating to the Internet.

FEDERAL NETWORKING COUNCIL (FNC) A council formed by the U.S. Office of Science and Technology Policy (an executive branch agency) to develop policy and provide direction for the National Research and Education Network. It coordinates the activities of different U.S. federal government agencies and was initially chaired by Dr. Charles Brownstein of the National Science Foundation.

FEDERAL STATE JOINT BOARD A board created by the FCC to regulate matters affecting state interests.

FEDERATION OF AMERICAN RESEARCH NETWORKS (FARNET) A nonprofit organization created in 1987 to advance the use of computer networks to improve education and research. It is headquartered in Waltham, MA.

FEEDBACK This refers to system messages that keep a user informed of system activities; the term also refers to data output from one machine that is input for another.

FEP Front End Processor. A device with stored logic which interfaces with data communications equipment to manage data traffic.

FF *see* Form Feed.

FIBER DISTRIBUTED DATA INTERFACE - *see* FDDI.

FIBER OPTIC CABLE A transmission medium that uses light waves to transmit data over glass fibers. This medium offers higher transmission rates and better security than electromagnetic signals but is more expensive.

FidoNet An informal coalition of bulletin board systems, begun in 1984, which are capable of exchanging messages and files. The name originally came from a bbs sites using Fido software, but as time progressed, a wide variety of bbs software packages have been supported.

FILE ACCESS DATA UNIT (FADU) A unit of information which can be handled and processed as an independent entity.

FILE SERVER A device on a LAN or other network that provides central storage for data files.

FILE TRANSFER, ACCESS AND MANAGEMENT *see* FTAM.

FILE TRANSFER PROTOCOL *see* FTP.

FILE TRANSFER PROTOCOLS Protocols define the rules or conventions that allow the transfer of data from one computer to another. They specify the amount of data to be sent in one continuous block, whether data may be sent in more than one direction at a time, error correction routines and when the transfer should be terminated. Examples include XMODEM, YMODEM, ZMODEM, etc.

FILTER A device which controls the frequencies which can pass along a data communications circuit.

FINGER PROTOCOL A TCP/IP protocol defined in RFC1288 (and earlier in RFC742) which allows the exchange of user information between hosts on the Internet.

FIPS An abbreviation for "Federal Information Processing Standard."

FIRSTSEARCH An information retrieval service developed by OCLC which has been designed with easy-to-use menus for the end user.

FIX *see* Federal Internet Exchange.

FIXED PATH PROTOCOL (FPP) In packet-switching networks this refers to a virtual circuit in which all packets relating to a particular session employ the identical path in the network.

FLAG A binary hex pattern used to mark the start and end of transmission frames.

FLAME A personal attack against the author of a message on the USENET or other electronic distribution list. The person doing the attacking is often called the "flamer."

FLOW CONTROL A process occurring in a data communications networks whereby the transfer of data between two points is regulated. Software flow control uses special control characters and escape sequences (e.g. XON/XOFF) to indicate when data transmission should pause to avoid losing characters. Hardware flow control is controlled by the presence of certain signals on the line (e.g. RTS/CTS request to send/clear to send).

FM *see* Frequency Modulation.

FNC *see* Federal Networking Council.

FNET The French branch of EUnet which was founded in 1983.

FOLLOWUP A response to a posting on a listserv, USENET or other electronic conferencing system.

FOREIGN EXCHANGE (FX) LINE A telecommunications circuit directly connected to a remote telephone exchange so that long distance charges are not made on each call to that LATA.

FORM FEED (FF) A special control code which increments the output of data to the top of the next form or screen, however it is defined.

FORWARD ERROR CORRECTION A type of error detection and correction using redundancy checking in which a large number of redundant bits are computed and added to the data stream so that if errors are detected, the receiving device can correct the errors without the sender having to retransmit the data.

FOUR WIRE CIRCUIT A communications line consisting of four wires constituting two 2-wire circuits. Each two-wire group carries signals in one direction with the resultant combination of bi-directional signals.

FPP *see* Fixed Path Protocol.

FRACTIONAL T1 A full T1 line transmits digital signals via time-division multiplexing over a 4-wire circuit through 24 logical subchannels which operate at 64 Kbps for an aggregate rate of 1.544 Mbps. For users who require the high speed and data integrity of T1, but cannot yet justify a full 24-channel line, carriers also often offer the use of one or more of the 64-Kbps subchannels.

FRAME A unit of information in transmittal which is bracketed by starting and ending flag sequences.

FRAME CHECK SEQUENCE A type of cyclic redundancy check sent with transmission frames in bit-oriented communication sequences. *see* Cyclic Redundancy Check.

FRAME CONTROL (FC) An ANSI FDDI and IEEE 802.x field.

FRAME FORMAT The structure of data within the indivisible unit (packet) of information for transmission. This may include originating address, destination address, check sequence, frame type and data.

FRAME RELAY One implementation of fast packet switching which uses a variable length unit of data called a frame. It uses statistical multiplexing over a shared network. Data comes into the switched network via bridges and routers using a frame relay interface. Once the data is within the switch fabric, the functions performed between nodes are minimal, thus speeding data transfer (for example, one third the number of intermediate tasks are required in a frame relay protocol as compared to an X.25 protocol). Frame relay appears to be a major player for packet networks at speeds of T1 and below while ATM is likely to dominate at speeds of T3 and above. Due to frame relay's growing popularity, many equipment providers are designing compatible interfaces and frame relay can support many data protocols including X.25, TCP/IP and SNA. *see also* Fast Packet Switching; Asynchronous Transfer Mode.

FRAME STATUS (FS) An ANSI FDDI or IEEE 802.5 field.

FRAMING BITS Extra bits which are added to data streams to provide control information in time division multiplexed circuits.

FrEdMail Free Educational Mail (FrEdMail). An informal telecommunications network which serves teachers and students in K-12 schools. It is made up of individual bulletin board systems (bbs), like FidoNet, which are operated by local institutions and sites may be customized to meet local needs. Gateways to regional networks (e.g. CERFnet in California) provide some connectivity to the larger Internet.

FREE EDUCATIONAL MAIL *see* FrEdMail.

FREE-NET A organization or system which offers a free community information system. These systems often provide access to the Internet. One of the largest of such networks is the Cleveland Free-Net which has been so successful that it has distributed its UNIX-based software to other communities around the United States. The Free-Net concept was originated by Tom Grundner at Case Western Reserve University in Cleveland, OH. The original Cleveland Free-Net also manages the National Public Telecomputing Network (NPTN) which acts as an information distribution arm of the Free-Net system (similar in concept to the National Public Radio).

FREEZE FRAME TELEVISION A method of transmitting multiple sequential video images. The image to be sent is frozen in a local memory prior to transmission.

FREQUENCY DIVISION MULTIPLEXING (FDM) Divides the bandwidth of a data channel into two or more subchannels on different frequencies, with a separation between the bands by a small band called a guardband. *see also* Multiplexing.

FREQUENCY MODULATION (FM) A technique to modulate a sine wave by modifying its frequency in order to carry information.

FREQUENCY SHIFT KEYING (FSK) A method of modulating a signal by using two different frequencies to distinguish between a mark (digital 1) and a space (digital 0) when transmitting on an analog line. This technique is normally used in low speed modems.

FRICC An abbreviation for "Federal Research Internet Coordinating Committee."

FRONT END PROCESSOR A computer which interfaces a host processor or a mainframe to a communications network.

FSK *see* Frequency Shift Keying.

FTAM File Transfer, Access, and Management. An application utility for file transfer in OSI that offers access to files on the network no matter what file format or host operating system. It supports the ability to open, close, read, write and delete files as well as the ability to access the whole file or parts of it. FTAM is not a database management system and no indexing or searching of files is supported (other than record keys) and only a single file may be opened at once.

FTP File Transfer Protocol. An application utility in TCP/IP as specified in RFC959 for the transferring of data files over the network. The protocol supports only the transfer of files, only complete files can be sent, and only a limited number of file structures are supported which include binary files, ASCII text files, EBCDIC text files and "paged" files. The "anonymous FTP" provision allows one to retrieve selected files from a remote system without prior authorization.

FTP/GOPHER GATEWAY PROJECT A project which allows FTP sites to be queried by a Gopher client.

FTPMAIL A service run by DEC to allow BITNET users who do not have direct access to the Internet to perform FTP file transfers from the Internet. It is a complementary service to BITFTP to help divide the traffic.

FULL DUPLEX (FD, FDX) Simultaneous independent transmission in two directions.

FULL TEXT DATABASE A database which contains the complete text of the original sources. These may include sources such as books, journals, directories, newspaper articles, court decisions, etc.

FUNCTIONAL PROFILE In the OSI standards this refers to the options selected to meet the requirements of the application program. Synonyms include functional standards or rifle shot standards.

FUNET The Finnish University Network was established in 1984 to provide network service to universities and research centers in Finland.

FX *see* Foreign Exchange.

FYI An abbreviation often used in electronic mail and news systems meaning "for your information." The term is also used to define a series of informative papers on the Internet.

GAIN The amplification of a message signal to compensate for signal attenuation.

GATEWAY A device connecting between two dissimilar networks that adds security, flow control, and protocol conversion. Gateways typically handle protocol-conversion operations across a wide spectrum of communications functions or layers. It requires software programming and central management. Gateways usually operate at the transport layer or above in the OSI model and provide protocol translation as well as routing. As a result of the more complex processing done in gateways they are usually slower in speed than bridges or routers.

GATEWAY-TO-GATEWAY PROTOCOL (GGP) A routing protocol used by the U.S. military for the Internet.

GBPS Gigabits per second.

GENERAL FORMAT IDENTIFIER (GFI) In the X.25 communications protocol, this is the first octet of a packet.

GES *see* Ground-Earth Station.

GFI *see* General Format Identifier.

G/G *see* Ground/Ground.

GGP *see* Gateway-to-Gateway Protocol.

GHZ *see* Gigahertz.

GIGABIT NETWORK A computer network which can transfer data at rates of one billion bits per second or more.

GIGAHERTZ One billion cycles per second.

GLOBAL ADDRESS The station address employed by the High-Level Data Link Control (HDLC) to send broadcast messages.

GNS NETWORK A commercial X.25 packet switching network run by BT North America. This network was formerly called TYMNET.

GO M-LINK *see* M-LINK.

GOPHER The Gopher is a distributed networking and document delivery system designed to work across any TCP/IP network. It was originally developed in 1991 by the University of Minnesota Microcomputer, Workstation, and Networks Center to assist campus users in accessing electronic information and navigating the Internet. The product was so well designed and simple to use, it was soon recognized as the campus-wide information system (CWIS) and has since been distributed and cloned at many other Internet sites around the world. The Gopher system has two major pieces, the server software and client software. The server accepts queries from users and responds by going to a menu, delivering a document, offering gateway access over the Internet, saving a file or printing a file. Users may access any Gopher server over the Internet in a "dumb" mode with a VT100 terminal emulation or by using client software which offers a graphical user interface which will provide windowing, pull-down menus, and other more elegant features. *see also* Gopher Development Team; TurboGopher; FTP/Gopher Gateway Project;

GOPHER DEVELOPMENT TEAM A group at the University of Minnesota that is actively designing and developing Gopher software.

GOSIP Government OSI Profile. An OSI based standard that the U.S. federal government has specified as an information processing standard for government contracts.

GOV Original high-level domain for non-military government organizations in the domain name system for the Internet (e.g. gpo.gov)

GOVERNMENT OSI PROFILE *see* GOSIP.

GREEN BOOK The remote login protocol developed for use in the JANET environment.

GREY BOOK The mail format and protocol used by the Joint Academic Network (JANET). It is similar to the protocols and formats used with SMTP. *see also* SMTP.

GROUND-EARTH STATION (GES) A satellite receiving site used for air/ground communication with mobile aircraft.

GROUND/GROUND (G/G) In aeronautics this refers to communications channels for fixed stations.

GROUP SEPARATOR (GS) A control character which is used to define or separate data into locally defined collections (groups).

GS *see* Group Separator.

GUI An abbreviation for "graphical user interface."

GulfNet A network serving Saudi Arabia and Kuwait.

HALF DUPLEX (HD, HDX) A circuit that permits data communications in two directions but not at the same time. Half duplex is non-simultaneous bi-directional communication.

HANDSHAKING The exchange of predetermined signals for the purposes of control between two devices.

HANGUP COMMAND This command disconnects a microcomputer and modem from the computer with which it is communicating.

HARDWARE FLOW CONTROL *see* Flow Control.

HARD WIRED In telecommunications a hard wired connection uses local cable, twisted wire or other direct connections of two nodes, stations or devices. In electronic circuitry, hard wiring refers to the performance of fixed logical operations due to the unalterable circuit layout rather than operations controlled by software.

HARNET The Hong Kong Academic and Research Network which was founded in 1986 to support higher education institutions.

HAYES COMMAND SET *see* AT Command Set.

HD *see* Half Duplex.

HDLC *see* High Level Data Link Control.

HDTV *see* High Definition Television.

HDX *see* Half Duplex.

HEADER The initial portion of a message which typically contains information such as the source or destination, length of message, priority, or time of origination.

HEANET Irish higher education authority network which was founded in 1985.

HEPnet High Energy Physics Network. This network serves physics research laboratories in the United States and Europe.

HERTZ (HZ) A unit of electromagnetic frequency equivalent to one cycle per second.

HIGH DEFINITION TELEVISION (HDTV) A television system standard in which a large number of lines are used to create the picture for higher resolution and clarity of the image. The implementation of HDTV will be key to the mass introduction of data services on a wide-scale basis into the home market.

HIGH LEVEL DATA LINK CONTROL (HDLC) A bit-oriented data link control standard developed by ISO which is similar to ANSI X3.66. The term is also used to refer to devices and the associated communications protocols by means of which checking is supported to allow that a transmission is successfully completed. In this standard, a message is transmitted in a series of frames that contain bits which are added for error control. These bits are checked to ensure that a message is properly transmitted and retransmitted, if necessary.

HIGH PASS FILTER A filter which allows all frequencies above a particular threshold to pass along a circuit. All lower frequencies are attenuated.

HIGH PERFORMANCE COMPUTING AND COMMUNICATION *see* HPPC.

HIGH SPEED MODEM Modems which operate at speeds of 9,600 baud and above.

HOST COMPUTER Mainframe and minicomputers are usually called hosts serving the needs of many users for central computing, networking and file sharing.

HPCA High Performance Computing Act. The name of the bill by former Senator Albert Gore (D-Tennessee) for S.1067 during the 101st Congress in 1990 to establish the National Research and Education Network (NREN). It was resubmitted in 1991 under S.272 by Gore and 18 cosponsors. At the same time, several other related bills were initiated in the 102nd Congress. H. 656 was passed by the House of Representatives on July 11, 1991 and two compromise bills were merged in the Senate and in November 1991 the compromised version of HPCA was passed by both the House and the Senate with President George Bush signing it into law on December 9, 1991. The bill signed by Bush authorized Congress to fund NREN but did not actually appropriate any money.

HPPC High Performance Computing and Communications. The name given to the program to develop computing, communications and software technology in the United States to meet its information and telecommunication needs. The HPCC Program will play a role to accelerate the development of the national information infrastructure (NII) which was outlined in the Clinton Administration's paper "Technology for American's Growth, A New Direction to Build Economic Strength," released on February 22, 1993.

HPPT *see* Hypertext Text Transfer Protocol.

HRC *see* Hybrid Ring Control.

HUB A central server in a star-configured network or the site at which branch nodes interconnect.

HUDX *see* Half Duplex.

HUM Unwanted electrical interference often caused by the alternating current power supply.

HYBRID RING CONTROL (HRC) An FDDI standard to provide control for integrated voice and data on a fiber optic local area network.

HYPERFTP A HyperCard stack that can be used by a Macintosh, using the MacTCP driver connected to the Internet, to preview text files on the FTP host before downloading.

HYPERMEDIA The ability to laterally go from one multimedia event (e.g. audio, still image, video, text) to another through a direct link by clicking on an icon or with some other technique.

HYPERLINK Links between words, phrases or multimedia icons in different documents, databases or systems.

HYPERTEXT The ability to laterally go from one document to another through a direct link based on keywords or phrases which have been dynamically connected via some type of interface.

HYPERTEXT TEXT TRANSFER PROTOCOL (HTTP) A public domain application layer protocol which uses TCP to transfer text over the Internet and is used for the design of information systems using Hypertext Links (Hyperlinks). Connections are usually made via the Telnet command with a specific Internet port (socket) being identified. World Wide Web uses an HTTP-like system for creating its links.

HYTELNET A hypertext utility for MS-DOS users (IBM PC compatibles), UNIX and VMS workstations which facilitates access to Internet resources. The product serves as a front-end to online library catalogs and other information resources on the network.

HZ *see* HERTZ.

IA5 *see* International Alphabet 5.

IAB Internet Architecture (Activities) Board. A group which makes decisions about standards and other important issues concerning the Internet. This group reports to the ISOC (The Internet Society). *see also* ISOC.

IASnet The Institute for Automated Systems Network. A data communications network for Socialist countries including Bulgaria, Hungary, former East Germany, Poland, Czechoslovakia, Cuba, Mongolia and Vietnam.

IBM CABLING SYSTEM The specifications for the types of cables to be used in connecting IBM products.

IBM LAN SERVER An IBM network operating system built around the LAN Manager.

IBM PC LAN An IBM network operating system built around MS-NET.

IBM TOKEN RING NETWORK A 2 Mbps LAN from IBM which uses token-passing techniques in a ring technology. Higher speed versions at 4 Mbps and 16 Mbps are also available (IEEE 802.5 is the standard).

IBM-VNET IBM corporate network.

ICD International Code Designator. A code assigned by the ISO as documented in ISO 6523.

ICMP An abbreviation for "Internet Control Message Protocol."

IDAPI *see* Integrated Database Application Programming Interface.

IDI *see* Initial Domain Identifier.

IDP *see* Initial Domain Part.

IDRP *see* Interdomain Routing Protocol.

IEE Institution of Electrical Engineers. A British-based society of professional engineers.

IEEE The Institute of Electrical and Electronic Engineers. A U.S.-based society of professional engineers.

IEEE 802.3 The protocol which specifies a CSMA/CD local area network. This is the standard on which the 10 Mbps Ethernet protocol is defined.

IEEE 802.4 The protocol which specifies a token bus local area network. *see also* Token Bus.

IEEE 802.5 The protocol which specifies the token ring local area network at speeds of 1, 4 or 16 Mbps.

IEEE 802.X A set of LAN standards developed by the Institute.

IETF Internet Engineering Task Force. A volunteer organization which updates the TCP/IP standards and investigates and solves technical problems on the Internet. It reports to the IAB. *see also* IAB.

IFLA International Federation of Library Associations and Institutions.

IGP *see* Interior Gateway Protocol.

IINREN Interagency Interim National Research and Education Network. The hierarchy of networks ranging from high-speed cross-country networks, to regional and mid-level networks, to state and campus network systems. The major federal components of the IINREN are the national research agency networks: NSF's NSFNET, DOE's Energy Sciences Network (ESnet), and NASA's NASA Science Internet (NSI). These agencies' networks constitute national network backbones that will collaborate in attaining NREN's gigabit speeds.

ILAN The Israeli Academic Network which is a branch of EARN.

ILLINET An automated library system consisting of two components: the Library Computer System (LCS) which has been operational as a statewide resources since 1980 and the Full Bibliographic Record (FBR) system. The system is based at the University of Illinois at Urbana-Champaign.

ILL PROTOCOL An interlibrary loan standard based on the OSI Reference Model that defines a standard for the interchange of ILL messages between systems which use dissimilar computers and software.

IMHO An abbreviation for "In my humble opinion."

IMODEM A file transfer protocol developed by John Friel for use with high-speed error-correcting modems. It offers no error detection or recovery routines and is considered a streaming protocol.

IMPLEMENTERS AGREEMENT A set of guidelines reached by private companies or service agencies who are developing hardware or software in support of a particular standard.

INFNET Istituto Nazionale Fisica Nucleare Network (National Institute for Nuclear Physics Network). The Italian nuclear physics network.

INFO GLOBE A major Canadian bibliographic information utility based in Toronto, Canada.

INFO-MAC A set of FTP archives on the Internet that is maintained at Stanford University and contains one of the largest collections of software for the Macintosh computer (FTP to sumex-aim.stanford.edu).

INFOPOP/WINDOWS A Windows-based client software package which offers a hypertext guide to the Internet with background, tutorials, and Internet destination information.

INFOPSI A DECNET network connecting academic and research institutions in Australia and New Zealand.

INFORMATION RETRIEVAL The general term used to describe the storage, searching and retrieval functions of a system in response to requests from users.

INIT MODEM The character string necessary to initialize a modem so that a communications session can begin.

INITIAL DOMAIN IDENTIFIER (IDI) The second field in the initial domain part of a network service access point (NSAP) address which identifies the numbering plan value (or addressing authority).

INITIAL DOMAIN PART (IDP) The first part of the network service access point (NSAP) address which consists of the AFI (authority and format identifier) and IDI (initial domain identifier).

INITIAL PROTOCOL IDENTIFIER (IPI) The first field in a protocol data unit (PDU) used for internal routing when a protocol selection technique is used.

INNOPAC The integrated library system developed, marketed and maintained by Innovative Interfaces Inc. (III). III is especially strong in the academic library market.

INSP An abbreviation for "Internet Name Server Protocol."

INTEGRATED DATABASE APPLICATION PROGRAMMING INTERFACE (IDAPI) An SQL API (application programming interface) developed by Borland based on the work of the SAG and X/Open consortia. *see also* SAG AND X/OPEN.

INTEGRATED SERVICES DIGITAL NETWORK (ISDN) A digitized telecommunications network being defined by CCITT in which data, voice, facsimile and video would be carried over the same communications channel using OSI standards. The first definition of ISDN published in 1984 provides a common user interface to digital communication networks on a worldwide basis. Access channels under definition include a basic rate defined by CCITT 2B + D (64K + 64K + 16K bps or 144K bps) and a primary rate that is DS1 (1.544 Mbps in the U.S., Japan, and Canada, and 2.048 Mbps in Europe).

INTEGRATED VOICE/DATA TERMINAL (IVDT) A unit which features a voice instrument, keyboard and video display. These are often designed to be used with a specific PBX on the customer premises or as part of some larger network.

INTEGRITY A property of data which ensures that it has not been corrupted in a computer system or network.

INTELLIGENT BUILDING A building in which a full range of connectivity, building environmental controls, phone service and other linked computers provide service to its occupants.

INTELLIGENT CONTROLLER A communications device which controls terminals or a cluster of devices for interacting with a network. These units may be programmable.

INTELLIGENT TERMINAL A terminal with built in memory and processing for performing more complex functions than a "dumb terminal."

INTERAGENCY INTERIM NREN *see* IINREN.

INTERCOM Internal communications within the same building or area but not outside the system.

INTERCONNECT A direct coupling of equipment whether it be acoustical, electrical or inductive in nature. This term is often used to denote connecting client equipment to the telephone network or the connection of other common carrier facilities to the telephone network.

INTERCONNECTION In computer networking this term defines a situation in which two computer systems can communicate but without consideration of how the interaction between the applications processes is controlled or how data are presented and recognized.

INTERDOMAIN ROUTING PROTOCOL (IDRP) A protocol for transferring policy-based routing information between Boundary Intermediate systems.

INTEREXCHANGE CARRIER (IXC) An authorized common carrier which passes information between exchange areas.

INTEREXCHANGE CHANNEL (IXC, IXT) A direct circuit between exchanges.

INTEREXCHANGE SERVICE *see* Long Distance Service.

INTERFACE Hardware, software and/or procedures for the transfer of information between interconnected equipment or processes.

INTERIOR GATEWAY PROTOCOL (IGP) An Internet U.S. military routing protocol.

INTERLOCATION TRUNKING Communications switching equipment which connects telecommunications channels from two or more different locations.

INTERMEDIATE SYSTEM (IS) An ISO system that supports the first three layers of protocols and provides connectivity between End Systems.

INTERMEDIATE SYSTEM HELLO (ISH) A signal (protocol data unit) which is sent by an intermediate system to announce its presence on a network.

INTERNATIONAL ALPHABET 5 (IA5) An international version of the ASCII character code with some special symbols (such as for different international currency).

INTERNATIONAL CODE DESIGNATOR *see* ICD.

INTERNATIONAL CONSULTATIVE COMMITTEE FOR TELEGRAPHY AND TELEPHONE *see* CCITT.

INTERNATIONAL DIALING PREFIX The special digits required to make an international telephone call.

INTERNATIONAL STANDARDS ORGANIZATION *see* ISO.

INTERNATIONAL TELECOMMUNICATIONS UNION (ITU) The telecommunications agency of the United Nations which develops standards for communications practices and procedures including worldwide radio frequency allocations..

INTERNET The collection of networks that connect government, university and commercial agencies (e.g. NSFNET, WestNet, etc.). The term is also more broadly used to designate any set of interconnected, logically independent networks. Many large internetworks, such as *the internet* are unified by the use of a single protocol suite, TCP/IP. A European example of an internetwork is IXI, an OSI-based network that interconnects many research networks throughout the European continent.

INTERNET ARCHITECTURE BOARD *see* IAB.

INTERNET PORTS Also known as Internet Sockets. These are extensions to a numerical IP address (or host/domain address) which directs a user to a particular application on a computer. Many times the Internet Port is defaulted (such as in many Telnet and FTP sessions) so that the user is not required to explicitly enter it.

INTERNET PROTOCOL (IP) Offers a common communications layer over dissimilar networks so that a packet of data may cross multiple networks on the way to its final destination. The protocol functions are described in MIL-STD 1777.

INTERNET RESOURCE GUIDE A guide published by the National Science Foundation (NSF) Network Service Center (NNSC) on resources available on the Internet.

INTERNET SOCIETY An organization devoted to promote the evolution, growth and use of the Internet on an international basis.

INTERNET SOCKETS *see* Internet Ports.

INTERNET TOUR HYPERCARD STACK A HyperCard stack entitled "Tour of the Internet" produced by The National Science Foundation (NSF) Network Service Center (NNSC).

INTERNETWORK PROTOCOL EXCHANGE (IPX) A subset of the XNS protocols used by Novell in NetWare.

INTEROPERABILITY The ability of devices from different companies to exchange information and to understand and operate on the data that are exchanged.

INTEROPERATOR A hardware device or software package which implements part of OSI standards and can work with hardware or software using other parts of the OSI model.

INTERPERSONAL MAIL SYSTEM (IPM) A part of the X.400 electronic mail standard.

INTERWORKING *see* Interoperability.

INTERWORKING UNIT (IWU) This term is synonymous with router, gateway or any intermediate system. It is a device which sits on a network and handles data traffic.

I/O Input/Output.

IP *see* Internet Protocol.

IP ADDRESS The address used to define the computer node number according to the TCP/IP protocol.

IPI *see* Initial Protocol Identifier.

IPM *see* Interpersonal Mail System.

IPX *see* Internetwork Protocol Exchange.

IRVING Project Irving is a network in the state of Colorado (initially funded with LSCA funds) to create a common interface and electronic link between disparate local online catalogs. It was meant to serve as a national model but only had limited success and is being superseded by Z39.50 interfaces.

IS *see* Intermediate System.

ISDN *see* Integrated Services Digital Network.

ISH *see* Intermediate System Hello.

ISO The International Standards Organization. An international group which promotes the development of computer (and other) standards. The OSI standards have been developed by this organization.

ISOC The Internet Society. The governing board to which the IAB reports. It is a membership organization whose members support the components of the international information network. *see also* IAB.

ISTC NETWORK A proposed high-speed national communications network in Canada proposed by *Industry, Science, and Technology Canada* in a 1990 feasibility study. It would be the Canadian equivalent of the National Research and Education Network (NREN) under development in the United States. The benefits of the network would include: improved competitiveness for industry, improved quality and productivity of research, more efficient use of scarce resources, greater opportunities for geographically remote organizations, and improvement of education.

ITESM Instituto Tecnologico y de Estudios Superiores de Monterrey. An academic research communications network in Mexico.

ITU *see* International Telecommunications Union.

IVDT *see* Integrated Voice/Data Terminal.

IWU *see* Interworking Unit.

IXC *see* Interexchange Carrier or Interexchange Channel.

IXI The Pilot International X.25 Infrastructure Backbone Service is designed to connect research networks in Western Europe to help reduce barriers caused by transborder data flow caused by public international telecommunications. It supports electronic mail, file transfer, information facilities and directory services. It is administered by the COSINE Policy Group, representing CEC national governments, and RARE (Reseaux Associes pour la Recherché Europeenne).

IXT *see* Interexchange Channel.

JABBER Noise on a network caused by the random transmission of data due to a malfunction.

JANET *see* Joint Academic Network.

JANET DNS The name registration scheme (NRS) for the Joint Academic Network (JANET). It the equivalent of the Internet Domain Name Server (DNS) scheme. *see also* Domain Name Server.

JITTER A signal impairment in which data are coming too fast or slow in the slots allocated in a circuit.

JOB ENTRY A set of computer instructions which allow the running of one or more programs in a batch mode.

JOB TRANSFER AND MANIPULATION (JTM) In OSI, an application layer protocol for initiating and controlling remote processes.

JOINT ACADEMIC NETWORK (JANET) A private wide area network which links the higher education institutions and selected research stations in the United Kingdom. Many libraries in the UK have their online catalogs available through this network. JANET is an X.25 packet-switched network and must be connected to the TCP/IP-based Internet through a gateway known as the "fat pipe."

JOINT ACCESS COSTS The fees associated with the telephone network which are a result of both local and long-distance use.

JTM *see* Job Transfer and Manipulation.

JUDDER The lack of uniform scanning on a telefacsimile picture.

JUNET Japan UNIX Network. A nationwide network that was created in Japan to support researchers and to act as a testbed for network experimentation.

JUGHEAD A Gopher tool which can be used to search the files of just one Gopher machine. It was developed at the University of Utah and is named after the "Archie" family of comic book characters.

JvNCnet North East Regional Research Network, formerly the John von Neumann Supercomputer Center Regional Network connected to the NSFNET. It serves primarily sites located in New Jersey, Pennsylvania, New York, Connecticut, Rhode Island, Massachusetts and New Hampshire. There are also connections in Colorado and Arizona.

K12NET A distributed microcomputer-based bulletin board system (bbs) network, which uses the FidoNet technology, to allow students and educators to do electronic conferencing on curriculum related issues. It was founded in 1990 and supports sites in North American, Australia, Europe and the Commonwealth of Independent States (formerly the USSR).

KBPS Kilobits per second.

KERMIT A microcomputer-based communications software package produced by Columbia University and widely used in academic institutions through site licenses. Kermit also offers a variety of file transfer protocols the more recent of which offer data compression, file attributes, sliding windows, and long blocks.

KEY MAPPING The ability to change the function of any of the keys on a keyboard. This is especially useful in communications software when needing to communicate with another information system.

KEYBOARD SEND/RECEIVE (KSR) A teleprinter terminal with a keyboard and printer for sending and receiving data. It has no data storage capabilities and data are transmitted on a character by character basis.

KHZ KiloHertz, 1000 Hertz or 1000 cycles per second.

KILOBITS A volume of data transfer which corresponds to 1,024 bits per second.

KILOHERTZ *see* KHZ.

KNOWBOT A experimental information retrieval technique in which a query can be sent around the Internet to find relevant information on a topic. They may be viewed as "robotic librarians."

KSR *see* Keyboard Send/Receive.

LADT *see* Local Area Data Transport.

LAN *see* Local Area Network.

LAN MANAGER A network operating system developed by 3Com and Microsoft that takes advantage of the OS/2 operating system. This term is also used to refer to a person who is responsible in an organization for the daily operations and maintenance of a LAN.

LAN SERVER A proprietary version of IBM's LAN Manager.

LANTASTIC A local area network (LAN) produced by Artisoft Inc. which incorporates CD-ROM networking capabilities into its basic architecture.

LAP An abbreviation for "Link Access Procedure."

LAPB *see* Link Access Procedure Balanced.

LAPD *see* Link Access Procedures Delta Channel.

LAPM *see* Link Access Protocol and MNP.

LAT Digital Equipment Corporation has developed this proprietary protocol for terminal services which provides communication for terminals across an Ethernet LAN.

LATA *see* Local Access and Transport Area.

LAYER The conceptual level of network processing functions. For example, in OSI Reference Model the network processing takes place in 7 different layers from the physical transmission of data up to the applications layer.

LAYER ENVELOPES In a communications protocol suite, a layered architecture is realized by enclosing information in envelopes. The software at each layer receives these envelopes from the layer above without altering it in any way but takes actions needed at the current layer and adds any necessary data of its own.

LAYER INDEPENDENCE The ability to modify the activities in any one layer of a communications protocol suite without affecting the layers above or below.

LAYER INTERFACE A means of communication between the different adjacent layers of the OSI model.

LAYERING The dividing up of a total communication system into a hierarchical set of smaller activities. For example: network or lower-layer protocols are those that deal with the issues of getting data from the originating system to a destination.

Application or upper-layer protocols are those which deal with the issues of allowing one computer system to effectively interoperate with another.

LCN *see* Logical Channel Number.

LCS Library Computer System. An automated library system originally developed by IBM for Ohio State University and later used by ILLINET.

LEAKY PBX A PBX that allows private-line traffic (tie-line) to access the local telephone exchange or vice versa.

LEASED LINE *see* Dedicated Line.

LEAST SIGNIFICANT BIT (LSB) The lowest valued bit within a field of bits.

LEC *see* Local Exchange Carrier.

LENGTH INDICATOR (LI) A field in a communications packet which indicates its length.

LF *see* Line Feed.

LI *see* Length Indicator.

LIBRARY COMPUTER SYSTEM *see* LCS.

LIBRARY AND INFORMATION TECHNOLOGY ASSOCIATION (LITA) A division of the American Library Association which focuses on library automation, information technology, telecommunications, networking and related issues.

LIBRARY SOFTWARE ARCHIVES *see* LIBSOFT.

LIBS A software program which uses the Art St. George OPAC (online public access catalogs) directory but runs only on UNIX and VMS systems. A menu system allows users to easily connect to the appropriate library system on the network.

LIBSOFT Library Software Archives. An FTP software archive site on the Internet located at the School of Library and Information Science at the University of Western Ontario which has one of the largest collections of library related software (FTP hydra.uwo.ca or 129.100.2.13).

LIBTEL A UNIX-based program which lists Internet accessible library catalogs and information systems on the Internet.

LINE A circuit connecting two or more devices or a communications path between two or more points including a satellite or microwave channel.

LINE DRIVER A device which boosts or conditions a signal between its place of origin and its destination.

LINE EXTENDER *see* LINE DRIVER.

LINE FEED (LF) A special control code which tells a peripheral device to increment the output line by one.

LINE SIDE CONNECTION A link to a telecommunications switch for the connection of terminals or other I/O devices.

LINE SPLITTER A device which divides a communications channel into two or more paths.

LINE TURNAROUND A reversal in the direction of a data signal when using a half duplex circuit.

LINE WRAP A setting in many telecommunications programs which indicates if a line of data should be wrapped to the next line when the incoming data has a greater column width than the screen can support (usually 80 characters).

LINK ACCESS PROCEDURE BALANCED (LAPB) In the X.25 packet switching protocol this refers to a standard link-level protocol.

LINK ACCESS PROCEDURES DELTA CHANNEL (LAPD) In the ISDN protocol this refers to a standard link-level protocol.

LINK ACCESS PROTOCOL AND MNP (LAPM) An error correction algorithm in data communications which detects data errors and retransmits bad data blocks. *see also* V.42.

LINKED SYSTEMS PROJECT (LSP) This is a development project by the Library of Congress, the Research Libraries Information Network, OCLC, and the Western Library Network to link their computer systems to exchange MARC authority records and MARC bibliographic records.

LINKED SYSTEMS PROTOCOL A national standard for bibliographic information retrieval developed by the National Information Standards Organization (NISO) and approved in 1988. The formal standard is Z39.50-1988 and is entitled "Information Retrieval Service Definition and Protocol Specification for Library Applications." *see also* Z39.50-1988.

LIS-LINK A mailbase group on JANET which is an electronic discussion group on computer-based information services, reference services, bibliographic instruction and networked services in the United Kingdom.

LISTSERV A software package originally developed by Eric Thomas for IBM BITNET host computers with the ability to manage electronic mailing lists. This software is widely used for computer conferencing and also has the ability to archive files and deliver them to users on demand. *see also* Computer Conference.

LITA *see* Library And Information Technology Association.

LLC *see* Logical Link Control.

LMS *see* Local Measured Service.

LOADING COIL An induction device used in local data communications loops (usually those longer than 5,000 meters) that compensates for wire capacitance and raises voice grade frequencies. These coils are sometimes removed when high speed data are being transmitted because of distortion. The term "loaded line" is also used.

LOCAL ACCESS AND TRANSPORT AREA (LATA) Geographic regions in the United States that define areas within which BOCs can offer local communications services (such as local phone calls, private lines, etc). After Divestiture there are approximately 160 exchanges called LATAs.

LOCAL AREA DATA TRANSPORT (LADT) LADT is based on the X.25 packet switching protocol and allows data to be sent over voice lines at higher frequencies than voice at speeds ranging from 1200 to 9600 bps.

LOCAL AREA NETWORK (LAN) A cluster of PCs and other computer peripherals in a relatively small area interconnected for the purpose of communications, file transfer and sharing of peripheral hardware.

LOCAL EXCHANGE A switching center in which a subscribers' line terminates. Calls made within the area supported by this local central office are usually considered local and not long distance.

LOCAL EXCHANGE CARRIER (LEC) A company which provides local exchange service.

LOCAL EXCHANGE SERVICE Telephone service within a local exchange area.

LOCAL LOOP A line linking a customers' telephone equipment to the local exchange (local central office).

LOCAL MEASURED SERVICE (LMS) A technique for costing phone or communications line charges based on time and or distance.

LOCALTALK Apple Computer's local area network using the AppleTalk network architecture and the network operating system called AppleShare.

LOCKING The ability to restrict or control access to files so that data are not destroyed when more than one person tries to simultaneously modify them.

LOG FILE A computer file which has been designated to receive data from a computer session.

LOGICAL CHANNEL The process of identifying each separate channel in a multiplexed circuit.

LOGICAL CHANNEL NUMBER (LCN) The number used to identify channels on an X.25 interface as well as between nodes on a subnetwork.

LOGICAL LINK CONTROL (LLC) A protocol developed for local area networks in standard IEEE 802 for data link-level transmission control. It is the upper sublayer of the IEEE layer 2 in the OSI protocol that complements the MAC protocol (IEEE 802.2).

LOGIN ID *see* User Login ID.

LOGOFF The process of terminating a connection to a computer system or data network.

LOGON The process of entering a computer system or data network.

LONG DISTANCE SERVICE Any telephone call made to a point outside the local service area of the originating station. The term "toll call" is also used.

LONG HAUL *see* Long Distance Service.

LONGITUDINAL REDUNDANCY CHECK (LRC) An error trapping technique in data communications in which a check character is captured at both the sending and receiving stations during the transmission. The check character is calculated with an algorithm using the odd and even parity of all the characters in the block.

LOOP A data communications configuration in which data links are connected in a series so that transmissions cross the loop in one direction. This is the same as a ring topology. *see also* Ring.

LOOPBACK TEST Also called message feedback. A test for a telecommunications channel in which the received data are returned to the sending end for comparison with the original data.

LOS NETTOS The Greater Los Angeles Area Network connected to the NSFNET.

LOW PASS FILTER A filter which allows all frequencies below a particular threshold to be transmitted on a circuit. All higher frequencies are attenuated.

LOWER LAYER PROTOCOLS Network or lower-layer protocols are those that deal with the issues of getting data from the originating system to a destination.

LRC *see* Longitudinal Redundancy Check.

LSAP An abbreviation for "link service access point."

LSB *see* Least Significant Bit.

LSP *see* Linked Systems Project; Linked Systems Protocol.

LU 6.2 Logical Unit 6.2. A software product from IBM that implements the session-layer conversation in IBM's SNA scheme in the Advanced Program-to-Program Communications (APPC) protocol.

MAASINFO A list of lists originally developed by Robert Maas to provide network users with a directory of lists for the Internet.

MAC Media Access Control. The lower sublayer of the data link layer as defined by IEEE 802 LAN standards. It control LAN traffic to avoid data collisions as packets move on and off a network through the adapter card (*see also* LLC, Access Protocol).

MAC-2 Medium Access Control-2. A new MAC protocol in which voice and data traffic are better integrated on an FDDI network.

MACHINE ADDRESS *see* NODE NAME.

MACHINE READABLE CATALOGING *see* MARC.

MAIL REFLECTOR A special automated electronic mail forwarding technique often used in implementing an email discussion group. A message is sent to the mail reflector and then is redistributed to all participants on a mailing list.

MAILBOX In the context of an electronic mail system, the mailbox serves as a storage location for messages waiting to be picked up by the recipient.

MAILER PROGRAMS A local electronic mail system such as OfficeVision (PROFS), ELM, Extended Berkeley Mailer, Z-mail, and VMS Mail.

MAJOR SYNCHRONIZATION POINT An OSI session layer service that pauses the flow of data until proper confirmation is received that previously sent data are stored.

MAN *see* Metropolitan Area Network.

MANAGEMENT INFORMATION BASE (MIB) A group of objects in a database which can be managed by the network management protocols.

MANCHESTER CODING A process of creating a self-synchronizing data stream through the merging of data and clock signals.

MAP/TOP Manufacturing Automation Protocol/Technical Office Protocol. MAP defines a series of OSI protocols and application utilities for the manufacturing environment and was originally developed by General Motors. TOP was developed by Boeing Corporation but applies to the office environment.

MARC MAchine Readable Cataloging. A standard developed by the Library of Congress and others to define the elements (fields) within a bibliographic record. The format has been expanded to provide a structure for other types of records (e.g. community information). The MARC standard was originally devised as a communications format and is the basis on which virtually all library systems share information and load records.

MARK The signal (communications channel state) which corresponds to the binary digit 1. The marking condition is present when current flows (in current loop channels) or when the voltage is more negative than -3 (in EIA RS-232-C channels).

MASSIVE OPEN SYSTEMS ENVIRONMENT STANDARDS *see* MOSES.

MAU *see* Medium Attachment Unit.

MBnet The Manitoba Network was founded in 1990 to server post-secondary institutions and other organizations involved in research. It is a mid-level network of CAnet in Canada.

MBPS Megabits per second.

MCI MCI Corporation the well-known long distance telephone carrier also provides Internet electronic mail access to its customers.

MDC *see* Mead Data Central.

MEAD DATA CENTRAL (MDC) A commercial timesharing information utility with an emphasis on offering full-text data through its LEXUS and NEXUS services.

MEAN TIME BETWEEN FAILURE (MTBF) The average time between consecutive failures on equipment or a network.

MEDIA ACCESS CONTROL *see* MAC.

MEDIUM The wiring or cable used to transport signals

MEDIUM ACCESS CONTROL-2 *see* MAC-2.

MEDIUM ATTACHMENT UNIT (MAU) A hardware device used in local area networks to attach the physical LAN wiring to computer equipment. It provides bit encoding and synchronization functions.

MEDLARS A major bibliographic utility operated by the U.S. National Library of Medicine with the major database being MEDLINE.

MEGAHERTZ (MHZ) One million cycles per second.

MELVYL One of the largest online public access catalogs in the world providing access to the library resources of the University of California. The system is managed and operated by the Division of Library Automation (DLA) of the University of California.

MERCURY PROJECT *see* Project Mercury.

MERIT The Michigan Educational Research Network connected to the NSFNET. It was founded in 1972 and is one of the oldest networks in the United States which serves the state of Michigan.

MERIT NETWORK *see* MichNet.

MESH NETWORK A network topology in which several nodes are interconnected and in some cases all nodes are connected to each other.

MESSAGE A unit of information which contains a complete thought.

MESSAGE HANDLING SERVICE *see* MHS.

MESSAGE STORE (MS) A service in the X.400 Message Handling Service (MHS) standard for the storing of electronic messages.

MESSAGE SWITCHING A method by which messages can be transferred between points not directly connected. This is usually done using store and forward techniques.

MESSAGE TRANSFER AGENT (MTA) In the X.400 MHS standard, this is the main processing element which stores and transfers email messages between User Agents and other MTA's.

META KEYS The ability to define "keyboard macros" to automatically execute a series of commands when an appropriate function key is depressed.

METROPOLITAN AREA NETWORK (MAN) A network which is too large to be treated as a single local area network but too small to be considered a wide area network. They are often under the control of a single administration.

MFEnet Magnetic Fusion Energy Network. This network was founded in 1975 and provides computer and supercomputer access in the United States and Japan to physics departments doing nuclear fusion research including all U.S. Department of Energy projects.

MHS Message Handling Service (System). MHS is a message transfer agent developed by Action Technologies Inc. and used by many commercial electronic mail packages. It supports the CCITT X.400 standard. It is the portion of an electronic mail system which delivers messages (may be text, graphics, spreadsheets, etc.) from one user or application to another. MHS is a store and forward architecture and provides a standard method of communication between disparate email packages, faxes, and gateway interfaces.

MHZ *see* MEGAHERTZ.

MIB *see* Management Information Base.

MichNet The statewide network in Michigan operated by Merit Inc. It was originally called the Merit Network but when Merit and IBM established a partnership to manage the NSFNET, the name was changed to MichNet to avoid confusion.

MICROCOM NETWORKING PROTOCOL (MNP) An error correcting technique advocated by the OSI reference model that resends lost or corrupt data in packets. MNP supports both interactive and file transfer applications over dial-up lines at rates up to 38.4 Kbps. It was developed by Microcom Inc. MNP Levels 1-4 enables error free asynchronous data transmission while MNP Level 5 incorporates the first four levels and also applies a data-compression algorithm. To use the MNP protocols both modems in a connection must use the same MNP protocols. Although MNP is proprietary, it became an industry standard in the 1980s because many users demanded it from manufacturers.

MICROWAVE A portion of the electromagnetic spectrum from about 760 MHz to 30 GHz.

MICROWAVE RADIO Microwave data links can easily carry up to 6,000 telephone circuits per channel but must have line-of-sight between towers (every 25-30 miles).

MIDDLEWARE A type of client/server software that assists in moving data between applications running on a computer and the network to which it is connected. The software provides standard APIs that save development time for programmers because they do not have to modify applications to accommodate network protocols.

MIDnet The Midwest Network connected to the NSFNET. It was founded in 1985 and links midwestern universities in Iowa, Illinois, Missouri, Arkansas, Oklahoma, Kansas and Nebraska.

MIL Original high-level domain for military organizations (e.g. army, navy) in the domain name system of the Internet (e.g. nic.ddn.mil).

MILNET One of the networks in the DDN (Defense Data Network) devoted to non-classified (non-secure) U.S. military communications. This network was built with the same protocols and technology as the ARPAnet but remained in place after ARPAnet was taken out of service.

MIME *see* Multipurpose Internet Mail Extensions.

MINOR SYNCHRONIZATION POINT An session layer service in the OSI protocol which does not require that the sending of data be paused to await confirmation.

MISC A newsgroup classification for material sent to USENET sites on miscellaneous topics. *see also* USENET.

M-LINK A service developed at the University of Michigan to allow public libraries to access and use resources on the Internet. Go M-Link is the Gopher front-end for the service.

MLP An abbreviation for "Multilink Procedure."

MNP *see* Microcom Networking Protocol.

MODEM A modulation/demodulation device which allows a digital device to be connected to an analog transmission network and vice versa.

MODEM ELIMINATOR A connector which interfaces between a local terminal (or other I/O device) expecting a modem and a computer expecting a modem so that the terminal and the computer can communicate. The device acts like an imitation modem.

MODEM7 An older file transfer protocol which is useful when sending data to and from older bulletin boards on computers which use the CP/M operating system.

MODULATOR A device which takes a signal and transforms it into another form suitable for transmission. A common transformation used in modems is to change analog signals into digital form and vice versa.

MOO GOPHER A Gopher client developed at Mankato State University which enables more than one person to view a Gopher listing and allows more than one user, through messaging, to instruct and lead one or more users through GopherSpace.

MORSE CODE A code developed in the 1830's by Samuel Morse in which numbers and letters are represented by various combinations of dots and dashes. In 1838, the code was first used over the electrical telegraph and it was first used on a commercial basis in 1844 for sending public telegraph messages between Baltimore and Washington.

MOSAIC A software package developed at the National Center for Supercomputing Applications at the University of Illinois which supports easy access to documents and graphics on the Internet. The software capitalizes on the World Wide Web product developed at CERN which enables uses to use hypertext links to jump to different documents on the Internet. The advantage of the Mosaic graphical user interface is that it uses hot links to photographs, video sequences, or other graphics and not just text. It was originally developed for the X-Windows operating system but copies for Apple Macintosh and Windows machines were later developed.

MOSAIC GRAPHICS SET A set of characters and control codes for those characters that are used for creating low-resolution graphic images on a videotex system.

MOSES Massive Open Systems Environment Standards. An organization of client/server users and manufacturers who cooperate to create client/server management standards.

MOST SIGNIFICANT BIT (MSB) The highest valued bit within a field of bits.

MOTIF A user interface based upon X-Windows developed by the Open Software Foundation which includes such vendors as IBM, HP, DEC and 40 others. It operates on a variety of UNIX platforms.

MOTIS An abbreviation for "message oriented text interchange system." *see also* X.400.

MRNet The Minnesota Regional Network connected to the NSFNET. It was founded in 1987 and connects a number of higher education institutions, businesses, and supercomputer centers in Minnesota.

MS *see* Message Store.

MS-DOS A popular operating system for microcomputers developed by Microsoft.

MS-NET A network operating systems developed by Microsoft which operates with MS-DOS 3.1 or later.

MSB *see* Most Significant Bit.

MSEN INC. A company in Ann Arbor, MI which provides Internet connections to the general public.

MTA *see* Message Transfer Agent.

MTBF *see* Mean Time Between Failure.

MULTIADDRESS CALLING A technique for broadcasting data to several stations in a network.

MULTIDROP LINE A single communications line with several interconnecting stations. Using this type of line often requires a polling mechanism. Also known as a multipoint line.

MULTIDROP NETWORK A network topology in which the various nodes or stations are stationed along a single transmission line.

MULTIPARTY EXTENSION A facility used in telephone services to support a conference call of three or more users.

MULTIPATH DISTORTION When signals arrive at slightly different times at a receiver due to different data paths.

MULTIPHASE MODULATION A type of phase modulation in which more than two alternatives of phase angles are used.

MULTIPLATTER A proprietary CD-ROM networking product produced by SilverPlatter, Inc. based on a Novell LAN. In 1993, SilverPlatter ceased distribution and maintenance on CD-ROM LAN products to concentrate on databasc and search engine development.

MULTIPLEXING A method for transmitting two or more separate channels of data on a single facility. Three techniques are commonly used: frequency-division multiplexing (FDM); time-division multiplexing (TDM); and statistical multiplexing.

MULTIPLEXER (MUX) A device which performs multiplexing.

MULTIPLEXOR An alternate spelling for multiplexer. However, it is used much less frequently in the literature.

MULTIPOINT A network consisting of more than two interconnected nodes or stations.

MULTIPOINT LINE *see* Multidrop Line.

MULTIPROTOCOL ROUTERS Routers which have the ability to handle data communications traffic using more than one protocol (e.g. TCP/IP, SNA, DECnet, etc.).

MULTIPURPOSE INTERNET MAIL EXTENSIONS (MIME) A format for handling and processing multimedia electronic mail messages (or composite text and multimedia formats) over a network. The MIME Internet standard is designed to be fully compatible with existing text-based email handlers.

MULTITASKING The ability to run more than one program at a time on a computer.

MUX *see* Multiplexer.

N-LAYER A generic reference to the "Nth" layer in a protocol suite.

N1 Inter-University Computer Network. The oldest academic communications network in Japan which is being superseded by the Science Information Network by NACSIS. *see also* NACSIS.

NACO National Coordinated Cataloging Operation, formerly the Name Authority Cooperative Project. A project begun in 1977 by the Library of Congress (LC) to create and support a national database of authority records. These name authority records are being transmitted via the Linked Systems Project.

NACSIS The National Center for Science Information Systems. This Center operates the Science Information Network to provide access to libraries, universities and research institutions throughout the Japanese islands.

NAK *see* Negative Acknowledgment.

NAME AUTHORITY COOPERATIVE - *see* NACO.

NAME REGISTRATION SCHEME (NRS) *see* JANET DNS.

NAMED PIPES An API for the OS/2 LAN Manager which allows the transfer of data from one application to another.

NAMESERVER *see* Domain Name Server.

NARROW BANDWIDTH A channel which carries signals in only a limited frequency range such as for voice.

NASA SCIENCE NETWORK *see* NSI.

NASA/RECON The major bibliographic utility operated by the U.S. National Aeronautics and Space Administration.

NATA North American Telecommunications Association.

NATIONAL COORDINATED CATALOGING PROGRAM *see* NCCP.

NATIONAL EXCHANGE CARRIERS ASSOCIATION (NECA) An organization created in 1983 by the FCC to watch over the collection, distribution and review of access charges to the interexchange carriers and customers.

NATIONAL PUBLIC TELECOMPUTING NETWORK (NPTN) An organization which provides electronic information distribution to local Free-Net systems operated around the United States. It was started by Tom Grundner of Case Western Reserve University as part of the Cleveland Free-Net. *see also* FREE-NET.

NATIONAL RESEARCH AND EDUCATION NETWORK (NREN) A proposed national electronic telecommunications infrastructure for the United States which would upgrade and expand the existing array of research networks. The network would support supercomputer users, libraries, database access, high performance computing tools, educational technology and specialized research facilities. Former Senator Albert Gore originally introduced the National High Performance Computer Technology Act of 1989 as S.1067 to enhance national competitiveness and productivity through a high-speed, high-quality network infrastructure supporting a broad set of services for research and instruction. The bill specifically asks that a 3 Gbps backbone be established in the United States.

NATIONAL TELEVISION SYSTEMS COMMITTEE (NTSC) A U.S. government committee which is responsible for the developing of standards for broadcast television in the United States.

NCA *see* Network Computing Architecture.

NCCF Network Communication Control Facility. IBM's host-based network management software that provides the operator interface and network logging facilities.

NCCP The National Coordinated Cataloging Program was begun in 1987 by the Library of Congress and the Association of Research Libraries to allow a small number of libraries to catalog at LC's highest quality level.

NCFTP A more user friendly version of the ftp program that supports automatic anonymous logins, clearer directory listings, percentage of sent messages (e.g. it indicates how much of an in-progress file has been sent, unlike ftp) and other features.

NCP PACKET SWITCHING INTERFACE *see* NPSI.

NCSAnet The National Center for Supercomputing Applications Network connected to the NSFNET. It was founded in 1986 and connects supercomputers to the NSFNET backbone in Indiana, Illinois and Wisconsin.

NCSU *see* Network Channel Service Unit.

NCTE *see* Network Channel Terminating Equipment.

NEARnet The New England Academic and Research Network connected to the NSFNET.

NECA *see* National Exchange Carriers Association.

NECTAR One of five gigabit network research testbeds for NREN. Located in Pennsylvania, this network concentrates on gigabit rate switch design and implementation.

NEGATIVE ACKNOWLEDGMENT (NAK) An indication that a previous transmission was flawed or in error.

NET Original high-level domain for network resources in the domain name system on the Internet; this is also an abbreviation for "Network Entity Title" which is the name of an active network device.

NETBIOS An API developed for IBM networks and is widely used as a standard interface to communications protocols. NetBIOS was originally developed as an interface between the IBM PC Network Program (superseded by PC LAN) and network interface cards provided by a company called Sytek. A grass roots movement of users is now pushing a combination of NetBIOS (operating at the OSI session layer) and TCP/IP. NetBIOS modules establish virtual communications sessions with each other across a network, giving the user the impression that they are directly connected with each other.

NETCOM A company which provides the general public with dial-up Internet connections and other communications services. It primarily serves people in California.

netILLINOIS The state of Illinois network connected to the NSFNET.

NETNORTH An academic and research network in Canada with connections to BITNET, NSFNET and the Internet. It was founded in 1984.

NETVIEW The network management product from IBM for its SNA networks. Data from non-SNA environments can be obtained through NetView/PC.

NETWARE Novell, Inc.'s network operating system.

NETWORK A linkage of computers or other devices by providing paths between users at a variety of geographical locations. Linkages between networks is possible through bridges, gateways and routers.

NETWORK ADAPTER The interface device that connects a workstation and file server to network cabling. These adapters contain electronic chips specifically designed for the network communication for the protocol required by the network operating system.

NETWORK ARCHITECTURE A definition of the logical components, functions and protocols on how a network should link and act.

NETWORK CHANNEL SERVICE UNIT (NCSU) A channel service unit (CSU) which is an on-premise piece of equipment for interfacing equipment to an ISDN network.

NETWORK CHANNEL TERMINATING EQUIPMENT (NCTE) A type of customer premises equipment (CPE) which converts the signals coming from a user's device into a form which can be handled by the data communications network.

NETWORK COMPUTING ARCHITECTURE (NCA) A proprietary communications protocol originally specified by Apollo Computer Inc. It has been designed to work in large networks with systems from multiple vendors.

NETWORK CONTROL The set of functions which manage the creation and dissolution of access paths between transmitting and receiving stations on a network. This type of control may either be done on a centralized basis (at one node) or on a decentralized basis.

NETWORK ETHICS The code of conduct which provides the guidelines, rules or backdrop for activities on a network. On the Internet two overriding principles include "individualism" and "the network is good and must be protected."

NETWORK FILE SYSTEM (NFS) Sun Microsystems, Inc. has developed this network service that allows transparent access to files located anywhere on the network. It is a set of protocols that allows one to use files on other network computers as if they were local. This includes the ability to read, write and edit files.

NETWORK INFORMATION CENTER (NIC) This term is used in a general sense to refer to any organization which provides management and training on or about a network. The Defense Data Network (DDN) NIC is active in the daily operations of the Internet through the performance of administrative functions such as the registering of IP addresses for host computers new to the network.

NETWORK INTERFACE CARD *see* Network Adapter.

NETWORK LAYER In the OSI Reference Model this is layer 3 which is responsible for routing data across the network. It handles the routing of data from the originating system to the destination system via any intermediate systems and makes it unnecessary for the upper layers to know the actual paths taken between the system.

NETWORK MANAGEMENT PROTOCOL (NMP) The protocol used in UNMA to communicate information from various vendors' element management systems to the Accumaster Integrating System. NMP is fully compliant with OSI.

NETWORK NEWS TRANSFER PROTOCOL (NNTP) This protocol is defined in RFC977 and specifies the distribution, inquiry and retrieval of news articles between TCP/IP sites on the Internet. NNTP allows articles to be stored in a central database and uses connection-oriented links (in real time) to allow the users' local host to access articles on the central host.

NETWORK OPERATING SYSTEM (NOS) The software that manages file sharing, security, electronic mail, methods for interfacing with different computer networks and any other resources of a network.

NETWORK OPERATIONS CENTER (NOC) The group responsible for the day-to-day management and operations of a communications network. On the Internet, each service provider has its own separate NOC so that one must know the proper organization to contact for problems.

NETWORK PROTOCOL DATA UNIT (NPDU) A packet of information as specified in the ISO Network Service.

NETWORK SERVER A central computing device on the LAN that provides for file storage, printing operations, modem connection, etc.

NETWORK SERVICE (NS) The service defined by ISO for offering either connectionless or connection-oriented networking.

NETWORK SERVICE ACCESS POINT (NSAP) The required addressing information to identify the user of a network service. It is the logical point of linkage between layer 3 (network service provider) and the user protocol in the OSI Reference Model.

NETWORK SERVICE DATA UNIT (NSDU) A unit of data which is transferred between peer network entities.

NETWORK TERMINATION NUMBER (NTN) The 10 lower digits in the X.121 addressing scheme.

NETWORK TOPOLOGY The physical layout of the nodes and interconnecting communication links which determine access paths through a network. The major topologies include the star network, loop (or ring) network, multidrop network, tree network, and mesh network.

NETWORK USER ADDRESS (NUA) A unique identification number given to a network user which identifies the address of a location from which the user originates. This address may also be used for billing or other purposes.

NETWORK USER IDENTIFICATION (NUI) A technique for identifying a user in an X.25 or other type of network.

NETWORKING AND TELECOMMUNICATIONS TASK FORCE (NTTF) A committee established by EDUCOM to address common interests in networks and telecommunications systems with an emphasis in the arena of higher education.

NevadaNet A mid-level regional network serving the state of Nevada with connections to NSFNET. It primarily serves institutions of higher education.

NEW ONLINE SYSTEM (NOS) *see* Prism.

NEWS A newsgroup category distributed to USENET sites which covers topics relating to network news and news reading software (e.g. news.groups).

NEWSGROUP On USENET this term refers to collections of articles or comments by readers on areas of common interest. A hierarchical classification technique is used to name discussion groups and the worldwide distribution of USENET newsgroups are divided into seven broad categories: comp, sci, misc, soc, talk, news and rec. Each of these categories is then subdivided into other subgroups by topic.

NEWSNET An online information utility which offers full test access to newsletters.

NFS *see* Network File System.

NIC *see* Network Information Center.

NII National Information Infrastructure. NII is part of the The High Performance Computing and Communications (HPPC) program initiated under the administration of President Bill Clinton to build the information and telecommunications infrastructure in the United States for the 21st century. *see also* HPPC.

NISC The SRI Network Information Systems Center maintans a database of RFCs, Internet drafts, and other information files. Many of these are made available via anonymous FTP and in print.

NLnet The Newfoundland and Labrador Network is a CAnet mid-level network in Canada. Founded in 1990, this network was designed to provide provincial data communications to support research, education and technology transfer in Newfoundland.

NMP *see* Network Management Protocol.

NNSC An abbreviation for the "NSF Network Service Center." BBN Systems and Technologies Corporation runs this center for the National Science Foundation. NNSC compiles *The Internet Resource Guide*.

NNTP *see* Network News Transfer Protocol.

NOC *see* Network Operations Center.

NODE A termination point for two or more communications links. In local area networks it also refers to a computer or single PC on a network.

NODE NAME A unique address for a computer connected to a communications network. This term is usually synonymous with "machine address."

NOISE The term given to interference which alters a data communications signal.

NONRETURN TO ZERO (NRZ) An encoding method used for signaling in which the binary representations of zero and one are indicated by a positive or negative voltage on a line. This method has no neutral or rest position.

NORDUnet The Nordic University Network which was founded in 1986 to provide network services for research and development in the Nordic countries including Sweden, Denmark, Norway and Finland.

NORMAL RESPONSE MODE (NRM) A multipoint primary/secondary operation for an HDLC mode.

NorthWestNet The Northwestern States Network connected to the NSFNET. Founded in 1987, this network provides service to 6 northwestern U.S. states including North Dakota, Montana, Idaho, Washington State, Oregon and Alaska.

NOS *see* Network Operating System.

NOTIS A major vendor for integrated library systems which is a wholly owned subsidiary of Ameritech Inc. The vendor is particularly successful in large academic libraries.

NPDA Network Problem Determination Application IBM's network management product that uses a series of panels to provide operators with information needed to perform problem determination and resolution.

NPDU *see* Network Protocol Data Unit.

NPI *see* Numbering Plan Indicator.

NPSI IBM X.25 NCP Packet Switching Interface. A software program developed by IBM which allows an SNA host computer to communicate with non-SNA DTE's (CCITT compliant) or other SNA hosts across an X.25 public data network.

NPTN *see* National Public Telecomputing Network.

NRCNET *see* CAnet.

NREN - *see* National Research and Education Network.

NRM *see* Normal Response Mode.

NRS *see* JANET DNS.

NRZ *see* Nonreturn to Zero.

NRZI An abbreviation for "nonreturn to zero inverted."

NS *see* Network Service.

NSAP *see* Network Service Access Point.

NSDU *see* Network Service Data Unit.

NSFNET A project begun by the National Science Foundation (NSF) in the mid-1980's to provide computer links between six supercomputers in the United States. This backbone has become the basis for the Internet which now connects thousands of local and regional data communications networks. The NSFNET is not the Internet but merely one of the many components of the international Internet. NSFNET has a number of purposes: to advance scientific collaboration on a national scale, to widen access to NSF funded supercomputers, to speed the dissemination of research results, to enhance education, to provide an experimental platform and to provide leadership in networking. The network has three levels of structure: a transcontinental backbone that connects a number of mid-level networks; mid-level networks consisting of regional, specific-discipline and supercomputing networks; and campus or organizational networks which link to the mid-level networks. In 1987, Merit Networking Inc. (Ann Arbor, MI) established a cooperative agreement with the NSF to re-engineer the network. During the next five years, Merit expanded and managed NSFNET and in 1992 it entered into a partnership with MCI and IBM to undertake further development.

NSI NASA Science Internet. The U.S. National Aeronautics and Space Administration's (NASA) agency network which is one of the backbone networks of the United States which will be one of the first networks to attain gigabit speeds in the NREN.

NSTN The Nova Scotia Technology Network was founded in 1990 and is a CAnet mid-level network serving Nova Scotia, Canada.

NTN *see* Network Termination Number.

NTSC *see* National Television Systems Committee.

NTTF *see* Networking and Telecommunications Task Force.

NUI *see* Network User Identification.

NULL ADDRESS A technique used by HDLC to support the transmission of commands which will be ignored by all secondary stations.

NULL CHARACTER An ASCII character that is used to fill a sequence of characters without affecting the meaning of the character string. However, these characters may affect control equipment on the network.

NUMBERING PLAN INDICATOR (NPI) CCITT has defined this 4-bit field, which is prefixed to X.121 and E.164 addressing schemes, and identifies the addressing authority. The is the first 4-bits of an 8-bit octet with the second set of bits being the type of address (TOA). *see also* Type of Address.

NUMERIC DATABASE A type of database which primarily contains numeric values from taken or summarized from original sources.

NWNet Northwestern States Network connected to the NSFNET.

NYSERNet The New York State Education and Research Network connected to the NSFNET. The network primarily serves New York state and was founded in 1985.

OARnet The Ohio Academic Resources Network connected to the NSFNET. Founded in 1987, this network connects most academic institutions in Ohio for the purpose of collaborative research, library systems and supercomputer access.

OCLC The Online Computer Library Center is located in Dublin, Ohio and is an international bibliographic utility used by libraries for cataloging, interlibrary loan, acquisitions, public databases and other related activities. Over 15,000 libraries from 40 countries are members.

OCR *see* Optical Character Recognition.

OCTET A set of 8 bits which make up a byte.

ODA *see* Office Document Architecture.

ODBC *see* Open Database Connectivity.

ODD PARITY A check done on a unit of data in which rows or columns of data add up to an odd number in order to allow the receiving device to determine if the data were properly transmitted.

ODIF An abbreviation for "Office Document Interchange Format."

OEM An abbreviation for "original equipment manufacturer."

OFFICE DOCUMENT ARCHITECTURE (ODA) A technique for specifying both the layout and content of an electronic document. Aside from supporting both text and graphics, the document may also contain audio and video components. ODA uses many elements from the Standard Generalized Markup Language (SGML).

OFFICEVISION IBM's proprietary electronic mail and calendar system. It replaces IBM's PROFS system. *see also* PROFS.

OFFLINE A condition which exists when a computer or piece of communications equipment is not being controlled by the host or is disconnected from the network.

OHIOLINK An effort to create an online system and electronically link the bibliographic resources of 19 academic libraries in the state of Ohio.

OLI *see* Open Link Interface.

OLIS *see* Oxford Library Information System.

OLTP *see* Online Transaction Processing.

1K-XMODEM-G A file transfer protocol which is a variation on the XMODEM protocol but was designed to be used with error-correcting modems. It is very efficient when used with the proper equipment. It sends and receives data on a continuous basis until instructed to stop. It does not allow batch transfers.

Onet The Ontario Regional Network was founded in 1988 to facilitate communications between educational and research institutions in Ontario, Canada. It is a mid-level CAnet network.

ONLINE Signifies a system in which end users are directly linked by circuits to a computer running an application.

ONLINE TRANSACTION PROCESSING (OLTP) A computer system which is optimized for the real-time monitoring, exchanging and updating of data in files. Examples of OLTP applications include booking airline reservations or circulation transactions in a library system.

OPEN DATABASE CONNECTIVITY (ODBC) An API developed by Microsoft for database connectivity in open systems using the SQL language.

OPEN LINK INTERFACE (OLI) An effort by Novell Inc. to establish a single interface for linking a variety of network adapter cards and networking protocols to its proprietary local area network and their NetWare operating system.

OPEN SHORTEST PATH FIRST (OSPF) A protocol used in TCP/IP.

OPEN SOFTWARE FOUNDATION A group of computer firms which is working on developing a standardized UNIX operating system.

OPEN SYSTEM A computing system that uses publicly available standards for its architecture. This allows easy development of third party products as well as easier networking with this type of system.

OPEN SYSTEM INTERCONNECTION *see* OSI.

OPERATING SYSTEM Software which interfaces the physical resources of a computer and an applications program.

OPTICAL CHARACTER RECOGNITION (OCR) A technique whereby the character content of scanned images may be converted into an appropriate character code (e.g. ASCII) to make the data available for textual manipulation in a computer system.

OPTICAL FIBERS *see* Fiber Optic Cable.

OPTICAL WAVEGUIDES A tubular optical transmission medium which carries data as light signals along a bounded medium made of glass fibers.

OPTI-NET A CD-ROM networking product which is NETBIOS compatible.

ORBIT SEARCH SERVICE A major bibliographic utility owned and operated by InfoPro Technologies (formerly Maxwell Online, Inc.).

ORIGINATING TRAFFIC A computer, terminal or other communication device which is responsible for putting data on a network.

ORG Original high-level domain for "other organizations" in the domain name system of the Internet (e.g. pac.carl.org).

OS/2 An operating system developed by Microsoft for IBM and its PS/2 line of computers.

OSAK OSI Applications Kernel. A full implementation of the OSI Session Layer in DECnet phase V that provides an API that manages the dialog between an applications program on an OSI network.

OSI Open Systems Interconnection. The general name used for ISO's effort to standardize computer-to-computer communications. The standard has a seven-layer model for communications and sections of the protocol are still in development. The OSI model defines an open system as one that obeys OSI standards in its communication with other systems. This contrasts with proprietary architectures (such as IBM's SNA) which are designed to support one vendor's equipment. The control levels in the OSI model include: layer 1-physical control layer; layer 2-link control; layer 3-network control layer; layer 4-transport layer; layer 5-session control layer; layer 6-presentation control layer; and layer 7-application layer.

OSIE An abbreviation for "open system interconnection environment."

OSI/NETWORK MANAGEMENT FORUM A group of companies which is focused on supporting network management standards for the OSI model.

OSPF *see* Open Shortest Path First.

OTHER COMMON CARRIER (OCC) A telecommunications company authorized by the FCC (other than AT&T) to provide communications services.

OVERFLOW A condition which exists when an existing circuit cannot handle all of the data traffic and the data must be rerouted in another path.

OVERLOAD *see* Traffic Overload.

OXFORD LIBRARY INFORMATION SYSTEM (OLIS) The integrated library system of the Oxford University Library System using the DOBIS/LIBIS system.

P/F An abbreviation for "poll/final."

PABX An abbreviation for "Private Automated Branch Exchange." *see also* PBX

PACCOM Pacific Communications Network. A fiber optic internetwork to support a group of telescopes in Hawaii and in the future Japan, Australia and New Zealand.

PACKET A unit of data which is transmitted at the network layer. It is also commonly used to denote an envelope of data bundled with addressing information for transmission through a network.

PACKET ASSEMBLER/DISASSEMBLER (PAD) A unit which takes character input and assembles it into packets (and disassembles at the other end) for transmission for a packet switching network such as with the X.25 packet switching network protocol. PADs are often used to support asynchronous dial-in access to X.25 communications networks.

PACKET DELAY A measure of how long it takes a packet of data to be transmitted from the originating source to the final destination in a packet switching network.

PACKET EXCHANGE PROTOCOL (PXP) A military communications protocol that is used on the Internet.

PACKET LAYER PROTOCOL (PLP) A name for a set of functions which allow the transfer of packets between two devices or stations.

PACKET RADIO A communications technique in which packets of data are sent via radio waves.

PACKET SEQUENCE In a packet switched network, each message is broken into individual packets which are routed through the network. These packets may not be sent via the same path or in the original order, but appropriate information is carried with each packet to reassemble the message in the original order.

PACKET SWITCH STREAM *see* PS.

PACKET SWITCHED NETWORK A communications network which carries data in the form of packets (e.g. X.25). The packet itself is internal to the particular network and any external interfaces may require conversion by an interface computer. It may be public or private.

PACKET SWITCHED PUBLIC DATA NETWORK (PSPDN) A public network that offers connectivity between users over an X.25 network.

PACKET-SWITCHING A method in data communications in which data are divided into packets and are routed to the final destination via the fastest route. The final destination node is responsible for reassembling the packets into the proper order.

PACKET SWITCHING EXCHANGE (PSE) A node in a packet switched network which has the ability to fulfill a full range of network activities such as the transfer of packets, assembling and disassembling of packets, and command functions such as the setup and disconnection of network calls.

PACKETIZING The process of splitting apart longer data streams and encapsulating them in packets so the data can be transferred over a packet switching network.

PACNET Pacific Network. An informal group of computer hosts throughout the Pacific region which link as an academic network. The primary countries involved include South Korea, Malaysia, Singapore, Indonesia, Australia and Hong Kong.

PAD *see* Packet Assembler/Disassembler.

PALS An integrated library system originally developed at Mankato State University (MN) and is the platform for statewide library networks in Minnesota, North Dakota and South Dakota. The product was originally marketed by UNISYS Corporation but in 1993 the marketing, maintenance and support was taken over by Dynix, Inc.

PAM *see* Pulse Amplitude Modulation.

PAR An abbreviation for "positive acknowledgment with retransmission" which is used in the TCP/IP environment.

PARALLEL TO SERIAL CONVERTER A device which converts a parallel transmission (i.e. data coming in over several signal paths at the same time) into a serial transmission in which the outgoing data are on one signal path.

PARALLEL TRANSMISSION The process of sending several bits (making up a byte) over separate signal paths at the same time from a computer to an external device. Parallel transmission can typically be done only over short distances. Contrast with Serial Transmission.

PARIS A packet network switching technology developed by IBM which uses Packetized Transfer Mode (PTM) to transmit variable length packets over SONET. A version of this product which supports the asynchronous transfer mode (ATM) technology is named PLAnet.

PARITY A constant state or equal value. Parity checking is an error checking routine in which character bit patterns are forced into parity (total number of one bits, odd or even) by adding a one or zero bit, as appropriate, as they are transmitted. The parity is then checked by the receiving device. If the sum of a series of bits is odd it is said to have odd parity, and if even then even parity.

PATH The complete set of resources necessary to enable a connection between two communicating devices.

PATH INDEPENDENT PROTOCOL (PIP) A technique used in packet-switching networks in which each independent packet can be routed independently of the other packets.

PATH INFORMATION UNITS (PIU) In the SNA networking environment, this refers to the name for layer 3 packets. The transmission header for these packets is generated by Path Control.

PATHNAME The address of a file or directory in a computer system. A full or "absolute" pathname specifies how to find a file or program from the root directory. A "relative" pathname indicates how to get to a file or program from the current working directory.

PBX Private Branch Exchange. A private switching system often located on a customer's premises. These may be automated (PABX) or manual.

PCI *see* Protocol Control Information; Presentation Context Identifier.

PCM *see* Pulse Code Modulation.

PDA *see* Personal Digital Assistant.

PDL *see* SPDL.

PDN *see* Public Data Network.

PDU *see* Protocol Data Unit.

PEAK LOAD The maximum traffic which is handled on a network.

PEER ENTITIES Entities within the same layer of a communications protocol.

PEER-TO-PEER NETWORK A network of independent computers which communicate and share resources. In this type of network, there is no single host computer through which all data must pass.

PEER-TO-PEER RESOURCE SHARING A software architecture that allows any node or station to contribute resources to the network while still running local programs.

PERFORMANCE SYSTEMS INTERNATIONAL *see* PSI.

PERMANENT VIRTUAL CIRCUIT (PVC) A permanent logical circuit in a packet switched environment (e.g. X.25) using packets and end-to-end control. Although many users are on a network at the same time, it appears to a terminal which is in a PVC that they have a dedicated connection to another computer. Once the circuit is established, the devices stay connected and no calls are necessary to establish the connection.

PERSONAL DIGITAL ASSISTANT (PDA) Small powerful computers, small enough to fit in coat pockets, which are designed to provide such features as the organizing of personal information, electronic mail, wireless networking, word processing, scheduling and other features. The product has been pioneered by Apple Computers.

PGI An abbreviation for "parameter group identifier."

PHASE INVERSION A technique used in phase modulation whereby the 0 and 1 bits are indicated by a change of signal phase in a particular direction.

PHASE JITTER A condition which exists when the length of signals are randomly distorted when the frequency of the signal changes.

PHASE MODULATION (PM) A technique for modifying sine waves for carrying information in which the carrier signal has its phase changed in accordance with the information to be transmitted.

PHASE SHIFT KEYING (PSK) A technique for transmitting digital data in which the carrier signal is modified to varying phases to represent data.

PHY-2 *see* Physical Layer Protocol 2.

PHYSICAL LAYER In the OSI model this is layer 1 and is responsible for the actual sending of the data on the transmission medium.

PI An abbreviation for "parameter identifier."

PICS *see* Protocol Implementation Conformance Statement.

PIP *see* Path Independent Protocol.

PIU *see* Path Information Units.

PLANET *see* PARIS.

PLASTIC FIBER A type of fiber optics in which the core is made of plastic and not glass. It is cheaper to produce but has a higher signal attenuation rate.

PLP *see* Packet Layer Protocol.

PM *see* Phase Modulation; Protocol Machine.

PMD *see* Private Management Domain.

POINT OF PRESENCE (POP) The particular location at which a data circuit terminates. In many networks, there may be many points of presence to indicate the major connection points for transceivers or other devices which provide network connections.

POINT-TO-MULTIPOINT VIDEOCONFERENCE An asymmetric conference in which all sites receive TV images but only one site can send them. The audio component may be bi-directional or unidirectional.

POINT-TO-POINT Data communications that occurs between two nodes on a network without passing through any intermediate nodes. The network layer handles this type of communications.

POINT-TO-POINT PROTOCOL (PPP) A protocol used to allow users to dial into the Internet (i.e. TCP/IP based network) with a high speed modem over a standard telephone line. PPP is a new standard replacing SLIP (serial line interface protocol) although PPP is less common but increasing in popularity. *see also* SLIP.

POINT-TO-POINT VIDEOCONFERENCE A conference between two sites, each of which can transmit and receive audio and video.

POLLING A means of controlling terminals on a multidrop line. Thus each terminal is sequentially queried to determine if any data are ready to send.

POP *see* Post Office Protocol; Point of Presence.

PORT An interface on a computer, terminal, network or other electronic device for the transferring of data. Also a point of access into a communications switch. The interface between a process and a communications or transmission facility.

POSIX A type of UNIX using IEEE standards.

POST OFFICE PROTOCOL (POP) This protocol specifies a process for the transfer of electronic mail with one POP machine being identified as the host and the other as the client. It is defined in RFC1225 and RFC1082 for a TCP/IP network such as the Internet.

POSTAL, TELEGRAPH AND TELEPHONE ORGANIZATION (PTT) This usually denotes a governmental department or agency that acts as a nation's common carrier in countries other than the United States or Canada.

POSTING An individual article or message sent to an electronic news group, listserv or electronic conferencing system.

POTS An abbreviation for "plain old telephone service."

PPDU An abbreviation for "presentation protocol data unit."

PPP *see* Point-to-Point Protocol.

PREFIX Any collection of data which precedes and qualifies a character stream which comes after it.

PREPnet The Pennsylvania Research & Economic Partnership Network connected to the NSFNET. Members include educational and research institutions, industrial firms and government agencies.

PRESENTATION CONTEXT In the OSI model this term refers to an abstract syntax and a selected transfer syntax.

PRESENTATION CONTEXT IDENTIFIER (PCI) A number associated with a particular presentation context.

PRESENTATION LAYER In the OSI model, this is layer 6 and is responsible for sending data in proper format between the application program and the lower layers. This layer supports screen display and is the home of special characters, control codes and special graphics for presentation to the user.

PRESENTATION SERVICE ACCESS POINT (PSAP) The logical point of connection between the presentation protocol and the user of the presentation service in an ISO compliant network.

PRESTEL The British Telecom public videotex service.

PRI *see* Primary Rate Interface.

PRIMARY A station on a network or application in a computer system which is given priority when contention occurs.

PRIMARY CHANNEL The main path for communications in a network as opposed to a secondary channel which may be used for diagnostic or backup purposes.

PRIMARY RATE INTERFACE (PRI) The collection of channels in ISDN used for the transfer of data and signaling information. The PRI has a bandwidth of 1.544 Mbps, using 23 B channels at 64 Kbps each, in the United States and Japan, the same as a T1 line. In Europe the PRI is rated at 2.048 Mbps since it uses 30 B channels and one D channel.

PRINT SERVER A computer on a network which makes one or more printers available to other users on the network.

PRISM The cataloging subsystem for the OCLC Online Computer Library Center was implemented in 1990.

PRIVATE BRANCH EXCHANGE *see* PBX.

PRIVATE LINE *see* Dedicated Line.

PRIVATE MANAGEMENT DOMAIN (PMD) A domain which is management by a private organization in an MHS network.

PRIVATE NETWORK A network exclusively setup and operated by or for a private organization, corporation, or individual.

PRO-CITE A bibliographic database management system software package developed by Personal Bibliographic Systems, Inc. (PBS).

PROCOMM A microcomputer-based telecommunications software package available either as public domain shareware or as a fully supported commercial product.

PRODIGY A videotex service offered in major cities in the United States.

PROFS A proprietary electronic mail and calendar system developed by IBM. It has been superseded by OfficeVision.

PROGRAM SUPPORT COMMUNICATIONS NETWORK (PSCN) A communications protocol suite which was developed for specialized scientific instruments. It uses a circuit-switched network for use on dedicated lines.

PROJECT IRVING *see* IRVING.

PROJECT MERCURY A project by Carnegie Mellon University (CMU), OCLC and Digital Equipment Corporation (DEC) to develop a state-of-the-art electronic library. It supports locally loaded databases, library catalogs, full-text files and electronic links with other libraries and information hosts. Z39.50 is used as the basic communications standard for linking UNIX workstations to the central information servers.

PROPAGATION TIME The time it takes for a signal to go from one end of a network to the other. This is also called propagation delay.

PROPRIETARY STANDARD A standard developed or owned by one company or group. Some companies license their proprietary products to third-party developers.

PROSPERO A distributed file system which provides tools to help users organize Internet resources. The Virtual System Model is the basis for Prospero and was designed and implemented by Clifford Neuman at the University of Washington.

PROTOCOL A collection of rules that control the exchange of information between the same layers in different nodes on a network.

PROTOCOL CONTROL INFORMATION (PCI) Header information which is added to user data for a protocol data unit (PDU) at a particular OSI layer.

PROTOCOL CONVERSION The process of changing the information used in one protocol into the format required by another protocol.

PROTOCOL DATA UNIT (PDU) A packet which consists of the control and data portion of a data unit.

PROTOCOL IMPLEMENTATION CONFORMANCE STATEMENT (PICS) A clearly identified set of functions which must be available for an application to be able to claim compliance for a particular implementation of a protocol. Many times, newer protocols clearly enumerate the minimum features necessary to claim conformance.

PROTOCOL MACHINE (PM) The real implementation of a communications protocol.

PROTOCOL STACK (PROTOCOL SUITE) A set of related protocols. This phrase is often used when referring to the protocols used for implementing the layers of the OSI model.

PS Packet Switch Stream. The name for British Telecom's public packet switching network.

PSAP *see* Presentation Service Access Point.

PSCN *see* Program Support Communications Network.

PSCnet The Pittsburgh Supercomputing Center Network connected to the NSFNET. It is an NSF supercomputing site and is a joint venture between Westinghouse, Carnegie Mellon University (CMU) and the University of Pittsburgh. It provides connectivity to the Internet through PREPnet, the Pennsylvania mid-level regional network.

PSE *see* Packet Switching Exchange.

PSI Performance Systems International is a company which provides individuals and companies with Internet access via dial-up or high-speed leased lines. It is located in Reston, VA. PSI runs a network called PSInet with nodes in more than 50 cities around the United States.

PSDN Packet Switched Data Network. *see* Packet Switched Network.

PSDS *see* Public Switched Digital Service.

PSK *see* Phase Shift Keying.

PSPDN *see* Packet Switched Public Data Network.

PSTN An abbreviation for" Public Switched Telephone Network (British);" "Public Switched Telecommunications Network."

PTT *see* Postal, Telegraph, and Telephone Organization.

PUBLIC DATA NETWORK (PDN) A packet switched network (e.g. X.25) which is available to the general public.

PUBLIC DIAL TELEPHONE NETWORK A communications network in which many users can do point-to-point connections using dial or push button telephones.

PUBLIC KEY An encryption technique in which one part of a decryption key is public and one part is secret.

PUBLIC NETWORK DEMARCATION POINT *see* Demarc

PUBLIC SWITCHED DIGITAL SERVICE (PSDS) A 56 Kbps full-duplex data network service offered by Bell Operating Companies.

PUBLIC SWITCHED TELECOMMUNICATIONS (TELEPHONE) NETWORK (PSTN) The public telephone network which uses packet switching technology.

PUBNET An alternate USENET category for newsgroups relating to information on public access UNIX systems (e.g. pubnet.sources).

PULSE AMPLITUDE MODULATION *see* Pulse Code Modulation.

PULSE CODE MODULATION (PCM) Transmission of information by modulation of an intermittent carrier. Pulse width, count, phase or amplitude may be varied. When the amplitude is varied it is called pulse amplitude modulation (PAM).

PVC *see* Permanent Virtual Circuit.

PXP *see* Packet Exchange Protocol.

QL SYSTEMS LIMITED A Canadian bibliographic utility.

QOS *see* Quality of Service.

QUALITY OF SERVICE (QOS) The acceptable level of communications service characteristics. A phrase used by the ISO for the 6 parameters that are defined for the communications environment offered for the exchange of applications data.

QUANTIZATION A technique used in pulse amplitude modulation (PAM), where an analog signal is converted into a digital form by sampling and the reassembly of the digital signal after transmission into an analog wave.

QUESTEL A major French bibliographic utility.

QUEUE In telephony this refers to phone calls being held or delayed while waiting for the line to be answered or a trunk line to become available. In data processing this refers to the sequencing of batch data processing sessions.

RACE Research and Development Program in Advanced Communications in Europe. This is a European project for the development of broadband technology.

RADIO FREQUENCIES The range of frequencies typically used in radio transmissions go from 30 Khz to 3 Ghz.

RADIO TRANSMISSION A type of telecommunications in which the signals are transmitted in the form of radio waves.

RANDOM NOISE A form of interference on a network which is not consistent. In telephone networks this term is used to refer to the white background hissing which often is audible.

RangKoM The Rangkaian Komputer Malaysia. Founded in 1987 this network links most universities in Malaysia.

RARP *see* Reverse Address Resolution Protocol.

RASCII A file transfer protocol which is similar to the ASCII file transfer protocol but performs no translation, filtering or flow control on the data being sent or received. It is a transparent protocol.

RBOC Regional Bell Operating Company. Also sometimes called Regional Bell Holding Company (RHC). Twenty two post-divestiture BOCs in the United States have organized themselves into seven regional Bell operating companies.

RCC *see* Regulated Common Carrier.

RD *see* Redirect; Routing Domain.

RDI *see* Routing Domain Identifier.

RDN An abbreviation for "relative distinguished name."

READY STATE A condition in which a DTE device (e.g. terminal) and DCE device are operable but not yet engaged in transferring data.

READY TO SEND A signal between a modem and its associated DTE (e.g. terminal) that the modem has made a connection with a remote device and that communication can begin.

REAL TIME An operational mode in which communications are immediately handled on an interactive basis.

REC A newsgroup category distributed to USENET sites which covers topics relating to recreational activities and hobbies (e.g. rec.food.drink).

RECEIVE LINE SIGNAL DETECTOR In EIA RS-232-C serial modems this is the signal which indicates to the attached terminal that it is receiving a signal from a distant source.

RECEIVE ONLY A device which is only capable of accepting or receiving data but not transmitting.

RECOGNIZED PRIVATE OPERATING AGENCY (RPOA) Private telecommunications carriers which provide public services.

RECORD MODE In computer communications this refers to the ability to log keystrokes or capture a data communications session.

RED RIDER A commercial microcomputer-based telecommunications software package for the Macintosh.

REDIRECT A type of protocol data unit (PDU) in the ES-IS protocol which tells an end system (ES) that a better route out of a subnetwork is available from another intermediate system (IS).

REDIRECTOR Software that reroutes network commands away from MS-DOS and onto the network in MS-NET.

REDUNDANCY CHECK An automatic or programmed check in which a systematic insertion of data is done and then verified at the receiving end to determine the degree of probability of error in transmission.

REFERENCE MODEL A high level description of a communications protocol suite.

REGENERATION In digital circuits a regenerator would detect, re-time and reconstruct the bits being transmitted.

REGIONAL BELL HOLDING COMPANY (RHC) *see* RBOC.

REGIONAL BELL OPERATING COMPANY *see* RBOC.

REGULATED COMMON CARRIER (RCC) A common carrier that is subject to federal regulation. For example AT&T as well as the BOCs are RCCs.

REJ An abbreviation for "reject."

RELAY To pass information on in a network or between networks such as is done with a router.

RELIABILITY OF SERVICE A measure of the "up-time" provided by a network or computer system. The term is also refers to the accuracy of data transfer.

RELIABLE TRANSFER SERVICE ELEMENT (RTSE) An ISO protocol used to transfer information between two application entities (AEs). It provides checkpoint and recovery mechanisms.

REMOTE FILE SERVICE (RFS) A distributed file system networking protocol developed by AT&T and is integral to UNIX V, Version 3 and is widely supported by UNIX market vendors.

REMOTE JOB ENTRY (RJE) The ability to transmit a job control stream for batch execution by a computer.

REMOTE LOGIN The ability of a user on one computer to connect and login to a computer somewhere else in a network. On the Internet this is done through the telnet command.

REMOTE OPERATIONS (RO) The ability for one computer or user to use another application on a remote system.

REMOTE OPERATIONS SERVICE ELEMENT *see* ROSE.

REMOTE PROCEDURE CALL (RPC) Calling a program or procedure from across the network.

REPEATER A device which boosts an electrical signal thus increasing the transmission distance possible.

REQUEST FOR COMMENTS *see* RFC.

REQUEST TO SEND (RTS) A control signal on a modem which indicates that it is OK to begin transmission.

RESEARCH LIBRARIES GROUP (RLG) An international library consortium headquartered in Stanford, California which among other things operates the Research Library Information Network (RLIN).

RESEARCH LIBRARY INFORMATION NETWORK *see* RLIN.

RESET The process of booting or reestablishing a communications session, communications device or computer system.

RESPONDER The end device in a communications network which responds to a communications request.

RESPONSE TIME The time gap between a terminal inquiry to a computer system and the receipt of a response.

RETRAINING *see* TRAINING.

REUNIR Reseau des Universites et de la Recherché (Network of Universities and Research). A metanetwork that connects French research institutions and universities.

REVERSE ADDRESS COMMUNICATIONS PROTOCOL (RARP) A U.S. military protocol which determines an IP address from a physical address.

REVERSE INTERRUPT (RVI) A control character sent by a receiving station to request the pause or termination of transmission.

RF FILTERS Radio frequency filters which are designed to accept radio waves at specified frequencies and to reject transmissions at all other frequencies.

RFC Request for Comments. In discussions on the Internet, this term refers to a set of documents (e.g. standards, idea statements) which have been published on the network.

RFS *see* Remote File Service.

RHC Regional Bell Holding Company. *see also* RBOC.

RI *see* Ring Indicator.

RICA The Andalusian research network that connects universities in Spain.

RING A network topology in which each node is connected in series to form a loop (contrast with BUS and STAR).

RING INDICATOR (RI) A signal used in modem interfaces defined under the EIA RS-232-C standard which indicates to the attached terminal equipment that an incoming call is present.

RIP *see* Routing Information Protocol.

RISER CABLE Wiring or cable used to connect communications devices on different floors in the same building.

RISQ The Reseau Interordinateur Scientifique Quebecois network was founded in 1989 to support the educational and research goals of al Quebec, Canada universities. It is a CAnet mid-level network serving Quebec, Canada.

RJE *see* Remote Job Entry.

RLG *See* Research Libraries Group.

RLIN Research Library Information Network. An international bibliographic information utility used primarily by libraries for cataloging, interlibrary loan, database access and other related functions. It's Citadel service provides libraries with online access to a variety of commercial databases.

RLOGIN A remote login protocol developed for UNIX systems.

RO *see* Remote Operations.

ROSE Remote Operations Service Element. An ISO protocol used to invoke remote operations between two entities. For example, the ability to invoke an application on a distant computer.

ROUTER A device which interconnects two networks. These function at the network layer of OSI and can determine the most efficient route by which to send data. They can segment packets of data and routers can detect and take care of loops in large networks (both of which bridges cannot do). Some disadvantages of routers are that they are normally slower than bridges and are usually protocol dependent so that, unlike a bridge, a specific router can only handle packets for OSI, TCP/IP, XNS, etc.

ROUTING The process of identifying the best path for sending data across a network.

ROUTING DOMAIN (RD) A logical communications domain which is centrally administered.

ROUTING DOMAIN IDENTIFIER (RDI) A unique code (identifier) which identifies each routing domain within one centrally administered network.

ROUTING INFORMATION PROTOCOL (RIP) A protocol used in TCP/IP networks which is based on the XNS protocol.

RPC *see* Remote Procedure Call.

RPOA *see* Recognized Private Operating Agency.

RS An abbreviation for "recommended standard."

RSCS Remote Spooling Control System. A proprietary IBM networking protocol used under the VM operating system.

RS-232-C A standard developed by EIA (Electronics Industries Association) which specifies a low-speed interface between computer equipment such as modems or printers and computers. It is ideal for data transmission over short distances (e.g. 50 feet) at speeds ranging from 0-20 Kbps.

RS-422-A A standard developed by EIA which specifies requirements for single-pathway communications between computers. This standard offers improved performance over RS-232-C connections.

RS-423-A A standard developed by EIA which specifies the electrical characteristics of unbalanced-voltage digital interface circuits.

RS-449 A standard developed by EIA which specifies a 37 pin interface and 9 position interface for data terminal equipment and data circuit terminating equipment employing serial binary data interchange.

RS-485 This standard resembles RS-422 except that associated drivers are tri-state, not dual-state. It may be used in multipoint applications sharing one central computer which controls many different devices (up to 64).

RS-530 This standard supersedes RS-449 and complements RS-232. It is based on a 25-pin connection used in conjunction with either electrical interface RS422 (balanced electrical circuits) or RS-423 (unbalanced electrical circuits). It defines the mechanical/electrical interfaces between DTEs and DCEs that transmit serial binary data. It supports higher data transmission rates than RS-232, however, RS-530 and RS-232 connections are not compatible.

RTFM An abbreviation for "Read the (......) manual."

RTG DOM An abbreviation for "routing domain."

RTS *see* Request to Send.

RTSE *see* Reliable Transfer Service Element.

RVI *see* Reverse Interrupt.

SA An abbreviation for "source address."

SAA Systems Application Architecture. IBM products use this architecture to provide a common a programming interface, a common means of user access and common communications support for IBM operating systems.

SABM *see* Set Asynchronous Balanced Mode.

SAG *see* SQL Access Group and X/Open.

SAP *see* Service Access Point.

SARM *see* Set Asynchronous Response Mode.

SAS An abbreviation for "single attached stations."

SASE Specific Application Service Elements. In OSI, those parts of the application layer which include FTAM, JTM and VT.

SCATTER/GOPHER An information retrieval paradigm in which documents are automatically clustered by subject and the user is presented with a table of contents of related items. The tool uses many of the interactive browsing features found in a server like "gopher" combined with more sophisticated searching tools from services like "WAIS."

SCI A newsgroup category distributed to USENET sites which covers topics relating to science and technology (e.g. sci.aeronautics).

SCREEN PAUSE The ability to freeze a screen while it is scrolling.

SCREEN SCROLL An option in telecommunications packages to determine if a line of data should be scrolled up one line when a carriage return is received and the screen is full.

SCREEN SNAPSHOT The ability to log the contents of the current screen to a disk file.

SCRIPT LANGUAGE A high-level communications programming language used in the development of communications applications.

SD *see* Start Delimiter.

SDLC *see* Synchronous Data Link Control.

SDN System Development Network. This network facilitates computer communications in the Republic of Korea.

SDNS An abbreviation for "secure data network system."

SDSCnet The San Diego Supercomputer Center Network connected to NSFNET.

SDU *see* Service Data Unit.

SE An abbreviation for "service element."

SEALINK A file transfer protocol similar to XMODEM except that it provides sliding window capabilities. It was developed by Tom Henderson of System Enhancement Associates in order to overcome problems with transmission delays caused by satellites or packet-switched networks.

SECONDARY In telecommunications networks this term often refers to nodes or stations which are designated as noncontrol (as with HDLC), in effect, one that can receive but not send commands. It may also refer to stations which are less critical in a network and can be worked around.

SECONDARY PROTOCOL IDENTIFIER (SPI) The second field in protocol data unit (PDU) which is used for internal routing in a network.

SEGMENTATION Chopping data streams into a smaller size for ease in transmission.

SELECTIVE REJECT (SREJ) An option used in HDLC satellite links to improve efficiency in asynchronous transmissions.

SELF TEST An internal test undertaken by a piece of equipment to check its ability to operate properly. For example, a modem may set up a loop back path through itself to test its transmitting and receiving function.

SEQUENCE A grouping of data elements into an order which makes sense.

SERIAL LINE IP *see* SLIP.

SERIAL TRANSMISSION The sending of data sequentially one bit at a time over a single channel. Modems for microcomputers usually operate in this mode.

SERVER A computer which provides clients with such services as databases, disk drives or access to other network resources. They can be mainframes, minicomputers, large workstations or LAN devices. It is possible for more than one server to supply services to clients. *see also* File Server.

SERVER MESSAGE BLOCK (SMB) A distributed network file system protocol developed by IBM and Microsoft to be used on the IBM PC LAN. This allows one computer to use the files and peripherals of another just as if they were local.

SERVICE ACCESS POINT (SAP) The boundary between different layers of the OSI model through which the different layers interact.

SERVICE DATA UNIT (SDU) An element of data which is passed to a service access point.

SERVICE PROVIDER In discussions on the Internet, this would refer to an organization which provides connections to the Internet.

Sesquinet The Texas Sesquicentennial Network connected to NSFNET. Founded in 1987, this network connects research sites in Texas to each other and to the NSFNET backbone.

SESSION A period during which a connection exists between two points in a network so that commands or data may be exchanged.

SESSION LAYER In the OSI model this is layer 5 and is responsible for directing user commands to a local operating system or to the network. At this level features are supported such as enabling two applications (or pieces of the same application) to communicate across the network, performing security, name recognition, logging, and administration.

SET A collection of data in which ordering is not important.

SET ASYNCHRONOUS BALANCED MODE (SABM) An operating mode command for the high level data link control (HDLC) protocol.

SET ASYNCHRONOUS RESPONSE MODE (SARM) An operating mode command for the high level data link control (HDLC) protocol.

SET NORMAL RESPONSE MODE *see* SNRM.

SFD *see* Start Frame Delimiter.

SGML Standard Generalized Markup Language. A device and process independent set of tags which are identified within the body of text for defining the structure of a document to indicate such features as headings, subheadings, paragraphs and body text.

SHARED TENANT SERVICE Several geographically closely related customers can be served by the same PBX equipment. Each customer is provided with separate attendant facilities, dedicated trunk lines and separate service requirements. This type of service is commonly offered to all the tenants in a building, research park, campus or the like.

(I.P.) SHARP ASSOCIATES LIMITED A Canadian online host specializing in numerical information in the financial, aviation and energy fields.

SHELL Software which operates on a workstation which transparently connects a user to the operating system or some application.

SHIELDED TWISTED PAIR Two insulated wires inside a larger cable which has been designed to reduce electromagnetic interference for outside sources.

SHORT HAUL MODEM *see* Short Haul Transmission Devices.

SHORT HAUL TRANSMISSION DEVICES This transmission medium is commonly 2 or 4 strand twisted pair wire which allows transmission from a few thousand feet to a few miles. The range depends on transmission rate, the transmission mode, and the transmitting and receiving devices.

SIGNAL An electrical impulse transmitted through a network to symbolize data, information or control.

SIGNAL-TO-NOISE RATIO The amount by which a signal exceeds its underlying noise.

SIGNALING The procedures involved with the establishment, maintenance and termination of calls on a network.

SIGNATURE In electronic mail this refers to the several lines of information that are included at the end of every message. This often includes such things as a person's name, email address, and institutional affiliation. Often people include quotations, pictures (usually made with ASCII characters) or other information.

SIMPLE INTERNET PROTOCOL (SIP) An Internet addressing protocol which is longer and more hierarchical for a ubiquitous and worldwide Internet addressing scheme.

SIMPLE MAIL TRANSFER PROTOCOL *see* SMTP.

SIMPLE NETWORK MANAGEMENT PROTOCOL see SNMP.

SIMPLE WIRE Twisted pair wiring used for telephone wiring inside residences or single-line business phones.

SIMPLEX CIRCUIT A circuit which allows the transmittal of data in only one direction at a time.

SIMTEL20 Simulation and Teleprocessing. An FTP software archive site on the Internet which contains public domain software (shareware) for MS-DOS, Macintosh, UNIX, CP/M and other operating systems. This archive site is maintained by the White Sands Missile Range (New Mexico) but is mirrored at some other sites around the country (FTP to WSMR-SIMTEL20.ARMY.MIL or 192.88.110.20).

SINGLE LINK PROCEDURES (SLP) A term used by CCITT which is synonymous with LAPB (Link Access Procedures Balanced).

SINGLE SIDEBAND TRANSMISSION A technique developed and perfected in the amateur radio community in which efficient use of a frequency can be made by filtering the carrier and the unwanted sideband of an amplitude modulated wave so that only the remaining sideband carries the information to be transmitted.

SIP *see* Simple Internet Protocol.

SLA An abbreviation for "Special Libraries Association."

SLP *see* Single Link Procedures.

SLIDING WINDOWS A technique used is some file transfer protocols in which information is sent at the same time it is being received, thus using less time waiting for the other modem to reply. Protocols using this technique are faster.

SLIP Serial Line Internet Protocol. A protocol that allows computer users at home to dial into a local Internet node with full Internet capabilities (e.g. telnet, FTP, email). SLIP is being superseded by PPP but is still very common. *see also* Point-to-Point Protocol.

SMAE The OSI System Management Application Entity.

SMASE An abbreviation for "system management application service element."

SMART MODEM A modem (modulator/demodulator) which contains sophisticated electronics to provide internal processing capabilities such as auto-answer, auto-dial, setting or changing the speed of transmission, or the ability to be programmed from a communications software package. Other capabilities often include internal error detection, data compression or the buffering of incoming or outgoing data.

SMARTCOM A commercial microcomputer-based telecommunications software package.

SMB *see* Server Message Block.

SMDS *see* Switched Multimegabit Data Service.

SMFA Specific Management Functional Area. In the OSI model this will define individual network management services.

SMI Structure of Management Information. Defines the basic format and identification of network management information in the OSI model.

SMILEY Smiling faces created with ASCII characters which are often used in electronic mail to indicate humor, anger or irony.

SMTP Simple Mail Transfer Protocol. An application utility in TCP/IP for electronic mail as specified in RFC821. It defines an envelope to be used for the delivery of electronic mail along with commands and conventions for performing the delivery.

SNA Systems Network Architecture. IBM's proprietary network architecture for communicating with IBM computers. Many non-IBM companies have developed products to support the SNA environment.

SNACP Subnetwork Access Protocol. A protocol used to access a subnetwork such as one running LAN or X.25 protocols.

SNAP Subnetwork Access Point. In the OSI protocol suite, this refers to the point of access between the internetworking protocol (layer 3b) and the subnetwork protocol (layer 3a). Several SNAPs may be resident at the layer boundary.

SNI Standard Network Interconnection. The mechanism that supports the intersystem communication for the Linked Systems Project. It follows OSI standards.

SNIFFER A network analysis tool in which one can catch and analyze packets of data on a network.

SNDCF Subnetwork-Dependent Convergence Function. The function which bridges across the differences in the Connectionless Network Service to the services available on the network.

SNDCP Subnetwork-Dependent Convergence Protocol. A protocol that adds services to the process of convergence between an SNAcP and an SNICP.

SNICP Subnetwork-Independent Convergence Protocol. A network protocol which is completely independent from the subnetworks which operate under it. The ISO CLNP is an example of an SNICP.

SNMP Simple Network Management Protocol. A TCP/IP network management protocol defined in RFC1157. It is supported by the Internet Activities Board (IAB) and is the network management protocol used by system administrators in the Internet.

SNOW REMOVAL The ability in some telecommunications software packages to eliminate flickering from a screen should it occur.

SNPA Subnetwork Point of Attachment. The point at which a computer system, subnetwork or internetworking unit is connected to an actual larger network.

SNRM Set Normal Response Mode. An operating mode command for HDLC.

SOC A newsgroup category distributed to USENET sites which covers topics relating to social issues and socializing (e.g. soc.singles).

SOCKETS A program which interfaces TCP/IP with Berkeley's version of UNIX.

SOCKETS, INTERNET *see* Internet Ports.

SOFTWARE FLOW CONTROL *see* Flow Control.

SONET *see* Synchronous Optical Network.

SPACE In telecommunications this refers to the absence of a signal. In telegraphy this indicates an open condition in which no current is flowing. In digital networks a space corresponds to a binary 0.

SPACE SEGMENT The portion of a satellite communications system between the ground transmitter and the satellite, not including any ground based links or equipment.

SPAG An abbreviations for "Standards Promotion and Applications Group."

SPAN Space Physics Analysis Network. Founded in 1981, this network is a NASA general purpose network serving institutions doing space related research largely in the U.S. with extensions to Canada, Japan and Europe.

SPDL Standard Page Description Language. The ISO standard for Page Description Language (PDL), such as PostScript, which defines the layout of an electronic document but does not specify the content. It specifies the how a page should appear when printed and tells the printer how to size and position fonts and graphics elements on a page.

SPDU An abbreviation for "session protocol data unit."

SPECIFIC APPLICATION SERVICE ELEMENTS *see* SASE.

SPECIFIC MANAGEMENT FUNCTIONAL AREA *see* SMFA.

SPECTRUM The range of electromagnetic frequencies.

SPF An abbreviation for "shortest path first."

SPI *see* Secondary Protocol Identifier.

SPIRES Stanford Public Information Retrieval System. An information retrieval and database management system developed at Stanford which has been licensed to over 40 other research, university and government institutions. Several libraries have taken the product and created their own integrated library system with SPIRES as the basis.

SPOOLING A method of keeping track of print jobs on a computer system and sending them to the appropriate printer, one at a time.

SPREAD SPECTRUM PACKET RADIO A packet radio technique, originally developed for military communications but now being used for implementing wireless LANs. Two techniques are commonly used: frequency hopping in which some data are transmitted on one frequency and then changing to another (this can also be coupled with time hopping); and direct sequencing in which the data are multiplied bit-by-bit with a code sequence before being transmitted.

SQL Structured Query Language. A common data description and access language for relational databases. The SQL language provides commands to describe, create, and delete tables of data used for relational databases, as well as to update the data within the tables. It was first implemented on IBM's DB2 in the early 1980s and has since become an industry standard with new versions such as ANSI SQL '89, ANSI SQL '92, and ANSI SQL3 (scheduled for release in the mid-1990s).

SQL ACCESS GROUP AND X/OPEN (SAG) SAG is a consortium of vendors in the database industry which is working on developing standards using client/server and other open system products. X/Open is another industry consortium focused on publishing and enforcing standards for open systems. Both groups are working together to establish an SQL standard and API's for client/server systems.

SREJ *see* Selective Reject.

SRI Stanford Research Institute. This organization operates a Network Information Systems Center (NISC) which plays an important role in the Internet.

SS An abbreviation for "session service."

SSAP An abbreviation for "session service access point."

ST An abbreviation for "straight tip."

STANDARD GENERALIZED MARKUP LANGUAGE *see* SGML.

STANDARD NETWORK INTERCONNECTION *see* SNI.

STAR A network topology in which a central device (Hub) has all other devices in the network directly linked to it. It is also possible to have several hubs linked forming a hierarchy.

STARLAN A local area network operating in a star or daisy chain topology and was originally developed by AT&T. It is now part of the IEEE 802.3 standard.

START BIT In asynchronous communications the start bit is longer than other bits and signals the beginning of transmission.

START DELIMITER (SD) A field in the ANSI FDDI and IEEE 802.5 communication protocols.

START FRAME DELIMITER (SFD) A field used in the IEEE 802.5 communications protocol.

STATION A location on a network with a unique address.

STATISTICAL MULTIPLEXING (STAT MUX) A multiplexing technique in which channel capacity is allocated based on the transmission of productive traffic. This technique takes advantage of the fact that data elements from two or more channels arrive in an asynchronous fashion.

STDM An abbreviation for "statistical time division multiplexing." *see* Statistical Multiplexing.

STE An abbreviation for "signaling terminal equipment."

STN INTERNATIONAL This cooperative online service is operated by Chemical Abstracts, STN-Karlsruhe, and the Japan Information Center of Science and Technology (JICST). The databases in the service are focused in chemistry and related fields.

STOP BIT In asynchronous communications the stop bit is longer than normal and indicates the end of a series of data bits.

STORE AND FORWARD NETWORKS A network in which messages are collected, packaged, transmitted, unpackaged and delivered by a network processor. Typically a system using this technique allows messages to be received and stored for retrieval or retransmittal at some later time.

STREAMS UNIX V.3's framework for network communications.

STRUCTURE OF MANAGEMENT INFORMATION *see* SMI.

STRUCTURED QUERY LANGUAGE *see* SQL.

SUBNET A network that is connected to a larger network via bridges, routers, or gateways.

SUBNETWORK ACCESS POINT *see* SNAP.

SUBNETWORK ACCESS PROTOCOLS *see* SNACP.

SUBNETWORK DEPENDENT CONVERGENCE FUNCTION *see* SNDCF.

SUBNETWORK DEPENDENT CONVERGENCE PROTOCOL *see* SNDCP.

SUBNETWORK INDEPENDENT CONVERGENCE PROTOCOL *see* SNICP.

SUBNETWORK POINT OF ATTACHMENT *see* SNPA.

SUBSCRIBER An organization or person who is registered as a valid user of a computer system, network or other information service.

SUN III *see* Sydney UNIX Network.

SUNET The Swedish University Network was established in 1980 to provide data communications between local and regional universities in Sweden. It is part of FUNET and NORDUnet.

SUPERNET The statewide data communications network for research institutions and academic sites in the state of Colorado. It provides Internet access to its users by connecting to WestNet, the regional mid-level network connected to NSFNET.

SURAnet The Southeastern Universities Research Association Network connected to the NSFNET. It was founded in 1987 and serves the southeastern U.S. states of Maryland, Virginia, Wet Virginia, Kentucky, Tennessee, Alabama, Mississippi, Louisiana, Florida, Georgia, South Carolina, North Carolina and the District of Columbia.

SURFnet Samenwerkingsorgaisatie Computerdienstverlening in het Hoger Onderwijs en Onderzoek (Cooperation Organization for Computer Services in Higher Educational Research). Founded in 1989, SURFnet provides networking for research and higher education institutions in The Netherlands.

SVC *see* Switched Virtual Circuit.

SWITCH An X.25 research network in Switzerland.

SWITCH FABRIC The technique for how data are switched from one node to another within a network rather than at the periphery of the network.

SWITCHED ACCESS The ability to use local switched telephone networks to reach interexchange networks. In computer networking this term refers to a network connection which can be created and destroyed as needed. This is the same as a switched virtual circuit.

SWITCHED MULTIMEGABIT DATA SERVICE (SMDS) SMDS is a networking technology targeted for public network applications. It offers a connectionless, public packet switching service known for high data throughput and low delay and error rates. It can be thought of as an attempt to provide LAN-like capabilities in a metropolitan-area environment. SMDS supports applications requiring a bandwidth of 1.2, 4, 10, 16, 25 and 34 Mbps. SMDS will be primarily used as an access technology and SMDS backbone networks will eventually be replaced by ATM backbones.

SWITCHED VIRTUAL CIRCUIT An X.25 network service which supports temporary network connections that are created and destroyed as needed.

SYDNEY UNIX NETWORK (SUN-III) A layered communications protocol suite which supports electronic mail, file transfer and remote printing.

SYNCHRONOUS DATA LINK CONTROL (SDLC) An IBM communications protocol associated with SNA. It supersedes BISYNC. SDLC offers full duplex transmission and is more efficient.

SYNCHRONOUS IDLE CHARACTER A control character used by synchronous transmission systems in the absence of any other characters to provide a signal.

SYNCHRONOUS OPTICAL NETWORK (SONET) A fiber optic carrier facility which is defined in several CCITT/ANSI standards for speeds from 51.84 Mbps to 2.5 Gbps. Individual channel rates are supported in 1.544 Mbps increments.

SYNCHRONOUS TRANSMISSION A transmitting device sends special characters (synch characters) at the beginning of an entire message and the message must be sent in sequence at a fixed rate. Contrast with asynchronous transmission.

SYNTAX The structure and format of data.

SYSOP *see* System Operator.

SYSTEM OPERATOR The person in charge of managing local functions on a computer.

SYSTEMS APPLICATION ARCHITECTURE *see* SAA.

SYSTEMS NETWORK ARCHITECTURE *see* SNA.

T1 A digital communications facility developed by AT&T for DS1 formatted digital signals that operates at a speed of 1.544 Mbps.

T1C A digital communications facility developed by AT&T for DS1C formatted signals that operates at a speed of 3.152 Mbps.

T2 A digital communications facility developed by AT&T for DS2 formatted digital signals that operates at 6.312 Mbps.

T3 A digital communications facility developed by AT&T for DS3 formatted digital signals that operates at speed of 44 Mbps.

T4 A digital communications facility developed by AT&T for DS4 formatted digital signals that operates at a speed of 273 Mbps.

T-CARRIER A series of transmission systems from AT&T using PCM at various channel capacities. *see* T1, T1C, T2, T3, T4.

TA *see* Terminal Adapter.

TAIL CIRCUIT An access line or feeder channel to a network node.

TALK A newsgroup category distributed to USENET sites which covers topics relating to long discussions on controversial topics (e.g. talk.abortion).

TALK PROTOCOL This protocol is used to control the real-time online interaction between two users on a network, similar to a written telephone conversation.

TANDEM NETWORK The connection of networks in a series so that the output of one circuit becomes the input of another circuit.

TANDEM SWITCH A switch in a tandem network for trunk connections.

TARIFF The published fees for a service, facility, equipment or other things offered by a common carrier.

TCM *see* Trellis Coded Modulation.

TCP/IP Transmission Control Protocol/Internet Protocol is a set of computer programs that enable communication between similar or dissimilar computers on a network. The Internet Protocol (IP) is the standard for sending basic units of data through an internet. TCP/IP enjoys wide vendor support particularly among universities and the federal government. It is the most widely used set of standardized, vendor-independent protocols and is the basis of the largest worldwide network, the Internet. TCP/IP can be carried over Ethernet, optical fiber links, DECNET, satellite links, and high-speed telephone lines. However, TCP/IP is not compliant with the emerging OSI standard which is universally expected to be the standard of the future. Thus migration paths will need to be developed to move from one protocol to the other.

TCSnet Thai Computer Science Network. Founded in 1988, TCSnet is a university network in Thailand.

TDM *see* Time Division Multiplexing.

TELECOMMUNICATIONS Communications between two or more parties using techniques to overcome physical barriers and/or distance.

TELECOMMUTING The process of making use of modern computer and communications technology to allow employers to work at home or some other site which is distant from the normal work site.

TELECONFERENCING A communications technique which allows a widely dispersed group to communicate simultaneously with each other.

TELEFACSIMILE *see* FAX.

TELEGRAPH SYSTEM A communications network which uses teletypwriters to transmit and receive messages. Telegraph networks have been in use for many years and traditionally have operated at very slow speeds (300 baud or less).

TELEGRAPHY Two way communications between sites using teletypwriters for ASCII text or telefacsimile for scanned documents.

TELENET A commercial packet switching network.

TELEMETRY A technique used for scientific investigation in which distant events are electronically monitored and the results are transmitted back to a receiving station for storage and eventual analysis.

TELEPHONE The general term used to describe the public switched telephone system. The term is also commonly used to refer to the equipment which connects to the system.

TELEPHONY The science and technology associated with the study and development of telephone networks.

TELEPROCESSING The integration of data communication techniques with data processing to create online real-time systems.

TELETEXT A general term for information services in which textual data is provided, often on television screens.

TELETYPE EXCHANGE *see* Telex.

TELETYPEWRITER A printer and a keyboard which is connected to a telegraph network. These devices have traditionally been very slow operating at speeds of 300 baud and below.

TELEX A dial-up telegraph service enabling uses to communicate directly between themselves by means of two-way circuits on a telegraph network. Western Union provides such services in the United States and overseas under the trademarks Telex and Telex II. The abbreviation means TELetype EXchange.

TELINK A file transfer protocol developed by Tom Jennings which is a variation on XMODEM but offers batch transfers and automatically transfers the file's name, size, and creation date attributes in the same session.

TELNET A software application utility for TCP/IP that provides terminal emulation and was specified in RFC854. The abbreviation stands for TELetype NETwork. This protocol allows users to login to other computers on the Internet.

10 Base T An IEEE standard which defines running Ethernet over twisted pair wires.

TENANT SERVICE *see* Shared Tenant Service.

TENET The Texas Education Network was founded in 1991 to serve as a data communications network for the Texas K-12 educational system. It provides access to the Internet through THEnet, the Texas Higher Education Network.

TEPI An abbreviation for "terminal endpoint identifier."

TERMINAL Hardware or equipment at the end of a communications circuit providing input/output. This term usually indicates a "dumb" terminal, microcomputer workstation with communications software, or a telephone set.

TERMINAL ADAPTER (TA) A protocol converter which offers X.25 devices access to an ISDN network.

TERMINAL EMULATION PROGRAM Software which allows microcomputers to emulate different terminal types. Different terminal emulations are often incorporated into popular communications packages.

TERMINAL SERVER A device that connects terminals (or other equipment) to a local area network. A server lets users of terminals connect to more than one computer.

TERMINAL-TO-HOST NETWORK A data network which connects "dumb terminals" (or microcomputers emulating "dumb terminals") to a central host computer. The majority of the processing done in this environment is done by the central computer. This type of network is often discussed using a master-slave metaphor.

TEXT FILE FORMAT A file which is stored in the ASCII (American Standard Code for Information Interchange) or EBCDIC (Extended Binary Coded Decimal Interchange Code) textual format.

TFTP An abbreviation for "trivial file transfer protocol."

THEnet The Texas Higher Education Network connected to NSFNET. Founded in 1984, this network connects academic, medical, research and corporations in Texas.

THICK ETHERNET A cable which contains copper wire that is both grounded and shielded. This cable is almost one-half inch thick in diameter and contains heavy shielding with polyvinyl and aluminum or copper.

THIN ETHERNET A version of Ethernet that operates on an RG-58 coax cable.

3+ 3Com's popular network operating system.

3270 The communications terminal standard developed by IBM used to communicate with IBM mainframes and compatible systems.

THROUGHPUT The amount of data transmitted through a network or system as a function of time.

TIE LINE *see* Interlocation Trunking.

TIME DIVISION MULTIPLEXING (TDM) Divides a data channel by time to support several subchannels. *see also* Multiplexing.

TIME SHARING A computer system or communications network which supports many concurrent users through separate processes.

TIME-TO-LIVE (TTL) The amount of time which passes from when an IP datagram is sent and it reaches its destination. A TTL field may be encoded in the datagram which serves as hop counter and is incremented in 1-second intervals. An ISO IP is measured in units of 500 msec rather than seconds.

TIMEFILL Characters (or bits) which are sent to maintain correct timing in synchronous transmissions. *see also* Synchronous Transmission.

TIMEOUT A set period of time after which a terminal or I/O device will perform some predetermined action (e.g. go back to a beginning menu screen).

TLI Transport Layer Interface. UNIX System 5 Release 3 incorporates this application program interface.

TLS *see* Transparent LAN Service.

TLV Type, Length, and Value. An encoding method used to designate a communications transfer syntax.

TN3270 A special variation of the Telnet program which allows users to properly interact with IBM mainframes.

TOA *see* Type of Address.

TOKEN A method of passing data bits in a network.

TOKEN BUS A local area network that uses the token passing method with a bus topology. The IEEE 802.4 is the standard for this topology and MAP specifies token bus as the LAN portion of its protocol stack.

TOKEN PASSING A method in which a token is passed around the network and is available to each station in the network. Only the station that has the token can transmit on the network (contrast with CSMA/CD). This is a form of time division multiplexing since only one station can transmit data at a time.

TOKEN RING A local area network that uses the token passing method as a ring topology.

TOP *see* MAP/TOP.

TOPOLOGY The physical layout of a network. It refers to the way in which the transmission media are interconnected to form a complete system. In LANs the major methods are star, bus and ring.

TOS *see* Type of Service.

TOUCH TONE *see* Dual Tone Multifrequency.

TP *see* Transport Protocol.

TP An abbreviation for "transaction processing."

TPDU *see* Transport Protocol Data Unit.

TP4 The transport layer protocol in OSI.

TR An abbreviation for "technical report."

TRAFFIC The capacity of data sent on a telecommunications system.

TRAFFIC ANALYSIS The monitoring, gathering and studying of data flowing across a network.

TRAFFIC OVERLOAD When data traffic flowing over a network exceeds system capacity and poor service for failure result.

TRAINING The procedure by which two modems make a connection in which they "discuss" and agree upon a data rate. Retraining is triggered by poor communication circuits in which both modems "agree" to lower communication speeds to compensate for line noise.

TRANSACTION A specific action on a data network in which there is an interaction between two systems to accomplish a particular function.

TRANSACTION LOG A record of all activities on a computer system or data network.

TRANSBORDER DATA FLOW The electronic transmission of data across national boundaries.

TRANSCEIVER A device that both transmits and receives. In an Ethernet system this unit provides a physical and electrical interface to the network cable and monitors the cable for activity to avoid data collisions. This device is also called a media access unit.

TRANSFER RATE OF INFORMATION BITS (TRIB) The rate at which data are accurately transmitted across a data link. TRIB rates are usually lower than actual transmission speeds from the originating computer because of error correction and other networking overhead factors.

TRANSFER SYNTAX The representation of data as it is sent across a network.

TRANSIT ROUTING DOMAIN (TRD) A term used in ISO standards which allows resources to be used by End Systems outside its domain to reach other End Systems in other routing domains.

TRANSMISSION CONTROL PROTOCOL/INTERNET PROTOCOL *see* TCP/IP.

TRANSMISSION ERRORS The change or alteration of signals in a transmission system caused by disturbances or distortion on the line.

TRANSMISSION MEDIUM The physical medium used to carry electronic signals (e.g. twisted pair wires).

TRANSPARENCY The ability to send binary data via character-oriented protocols. This may include control information as well as content.

TRANSPARENT In networking this term indicates connections which actually exist but are not visible to the user. It is invisible.

TRANSPARENT LAN SERVICE (TLS) A product from U.S. West which extends a local area network infrastructure over an entire metropolitan WAN area. Using fiber optic cable the service can link as many as seven LAN locations within a twenty five mile fiber loop. It can support native LAN speeds of 4, 10 or 16 Mbps.

TRANSPONDER A device which receives a transmission signal and retransmits it after amplification. These are commonly used on satellites.

TRANSPORT LAYER In OSI this is layer 4 and is responsible for keeping end-to-end communications. If the network crashes, the transport layer software will look for alternate routes or save the transmitted data until the network connection is reestablished. Error correction is provided at this level.

TRANSPORT PROTOCOL (TP) This protocol is layer 4 (ISO 8073) of the OSI Reference Model which supports five classes of operation based on the Network Service Provider. The U.S. military standard which is the equivalent of this is the Transmission Control Protocol (TCP) which is specified in MIL-STD 1778 and is used on the Internet.

TRANSPORT PROTOCOL DATA UNIT (TPDU) The unit of data which is sent in the Transport Protocol.

TRANSPORT SERVICE ACCESS POINT (TSAP) The logical point of connection between the Transport Service Provider and the user.

TRD *see* Transit Routing Domain.

TRELLIS CODED MODULATION (TCM) A technique whereby several characteristics of a signal are changed to achieve forward error correction. This technique is used in CCITT V.32 and V.32bis compliant modems.

TRIB *see* Transfer Rate of Information Bits.

TRIPLE-X A set of three CCITT recommendations which permit remote login over X.25 networks. The three standards included under this general heading include X.3, X.28 and X.29. *see also* X.3; X.28; X.29.

TRUNK A transmission path to interconnect the exchanges in the main telephone network. It is the telecommunications circuit between the nodes of two networks. The term also refers to a telephone exchange line that terminates at a PBX.

TS An abbreviation for "transport service."

TSAP *see* Transport Service Access Point.

TSDU An abbreviation for "transport service data unit."

TTL *see* Time-to-Live.

TURBOGOPHER Macintosh Gopher client software.

TURNAROUND TIME The time necessary to reverse the direction of a transmission. For example the time to change from a receive mode to a transmit mode in a half duplex line.

TWA *see* Two-Way Alternate.

TWISTED PAIR A wiring technique in which two insulated wires are wrapped around each other but are not covered with an outer sheath. It is most commonly used in phone systems but is a low-cost medium for LAN connections or directly connecting workstations to a host computer. Two kinds are available: unshielded (as used with the phone system) and shielded which provides some limited protection against interference.

TWO-WAY ALTERNATE (TWA) A technique whereby two stations (or protocol machines) alternate communications flow. This often happens for link control processes but can apply to higher-layer protocols too.

TWO-WAY SIMULTANEOUS (TWS) The ability for a network and two communicating devices to send and receive data at the same time. *see also* Full Duplex.

TWO WIRE CHANNEL A circuit with two individual wires in a pair (or the logical equivalent) for half duplex transmission.

TWO WIRE CIRCUIT A circuit formed using twisted pair wiring that can be used for either half or full duplex operation.

TWS *see* Two-Way Simultaneous.

TWX An old term which refers to Western Unions Telex II service. *See* Telex.

TYMNET The old name for a widely used X.25 commercial packet switching network. It is now run by BT North America and is called the GNS network.

TYPE OF ADDRESS (TOA) CCITT has defined this 4-bit field for indicating the type of address contained in the concatenated address fields that follow. It is the second half of an 8-bit octet, of which the first half is the NPI (numbering plan indicator). *see also* Numbering Plan Indicator.

TYPE OF SERVICE (TOS) U.S. Military communications protocols use this field to indicate the logical address assignment for the upper layer use of the protocol.

UA *see* User Agent.

UDP User Datagram Protocol. One of the basic protocols used on the Internet which was originally developed by the military to provide the equivalent of an ISO connectionless transport.

UI *see* Unnumbered Information.

UKnet The British portion of EUnet.

ULP *see* Upper Layer Protocol.

ULTRIX A version of UNIX developed by Digital Equipment Corporation.

UNA Universitats-Netz Austria (Austrian University Network). The network connects Digital computers at Austrian Universities. The network may be merging with ACONet.

UNAM Universidad Nacional Autonoma de Mexico. A network which links the various sites of the National University in Mexico City.

UNBOUNDED MEDIUM A transmission system in which the signal is radiated in many directions rather than along a bounded path (e.g. fiber cable).

UNBUNDLED Services, training or equipment that can be sold separately.

UNCOVER A table-of-contents indexing service which provides electronic access to almost 20,000 journal titles as well as document delivery for those titles. Most articles which are delivered through UnCover are scanned and stored on optical media for quick access by subsequent requesters. UnCover is owned an operated by UnCover Company which is a wholly owned subsidiary of CARL Systems, Inc. and B.H. Blackwell's. *see* CARL Systems, Inc.

UNI An abbreviation for "User Network Interface."

UNIFIED NETWORK MANAGEMENT ARCHITECTURE *see* UNMA.

UNIFORM RESOURCE LOCATOR (URL) A standardized technique to point to information resources (i.e. applications) on the Internet. This standard pointer format, known as the URL, points to a file on the network, but if the file is moved, access is lost. To solve this problem, the Internet Engineering Task Force has developed another convention using Uniform Resource Names (URN). *see also* Uniform Resource Names.

UNIFORM (UNIVERSAL) RESOURCE NAMES (URN) A standardized naming convention developed by the Internet Engineering Task Force (IETF) to symbolically name information resources on the Internet. Servers on the network are required which will tell where applications reside on the Internet. *see also* Uniform Resource Locator.

UNInet A university computer network in Indonesia which was founded in 1988.

UNINETT The University Network was established in 1978 to manage and connect all higher education institutions in Norway.

UNITE User Network Interface to Everything discussion list on the Internet. The purpose of this list is to act as a discussion focus related to one interface for all network services.

UNIX One of the most popular multi-user and multitasking operating systems which was originally developed by AT&T. There are many different versions of this operating system and it is easily ported to different kinds of computers. Although UNIX systems have played an important part on the development of the Internet, they are not necessary to use the Internet.

UNIX TO UNIX COPY PROTOCOL *see* UUCP.

UNMA Unified Network Management Architecture. A proprietary three tiered architecture for network management developed by AT&T. It is an open architecture which employs interfaces based on OSI standards.

UNNUMBERED INFORMATION (UI) A frame format which has been specified for HDLC and LLC to provide connectionless data transfer services.

UNSUBSCRIBE The ability to stop receiving messages from a LISTSERV or other electronic distribution facility.

UPLINK The portion of a satellite circuit going from the ground station to the satellite.

UPPER LAYER PROTOCOL (ULP) Application or upper-layer protocols are those which deal with the issues of allowing one computer system to effectively interoperate with another. This general expression usually refers to service-oriented protocols above layer 4 in the OSI model.

URL *see* Uniform Resource Locator.

URN *see* Uniform Resource Name.

USAN The National Center for Atmospheric Research's (NCAR) University Satellite Network. Founded in 1987, this network connects institutions to the National Center for Atmospheric Research (NCAR) via satellite links.

USENET User's Network. A worldwide network of over 10,000 hosts and 300,000 users which supports newsgroups -- informal discussion groups on topics of common interest. There are hundreds of newsgroups and much of the USENET traffic is now carried on the Internet. However, USENET predates the Internet and has users on five continents with the highest concentration of users in the United States. *see also* UUNET.

USENET ADDRESS SERVER A database of Usenet participants for anyone who has posted a message to a USENET newsgroup.

USER AGENT (UA) A process which manages and controls the services offered on a network by some application service element (e.g. the MHS Directory Service).

USER DATAGRAM PROTOCOL *see* UDP.

USER LOGIN ID A unique code which identifies a user who is trying to access a network or computer system.

UTLAS A Canadian bibliographic utility which supports libraries in services such as cataloging, interlibrary loan, retrospective conversion, etc. at its central computer center in Toronto, Canada.

UTP An abbreviation for "unshielded twisted pair."

UUCP UNIX-to-UNIX Copy Protocol. A software facility for copying files between UNIX systems. This tool provides the basis on which email and USENET services were built in the UNIX environment.

UUNET A nonprofit telecommunications organization which provides USENET users with mail, news and low-cost contracts through volume discounts. Uunet Technologies runs AlterNet, a public TCP/IP network service offering Internet and commercial network services. *see also* USENET.

V SERIES The collection of CCITT standards and definitions that cover data transmission over telephone networks.

V.1 The CCITT standard which specifies the equivalence between binary notation and the significance of a two-condition code.

V.2 The CCITT standard which specifies the necessary power levels for signal transmission on a telephone circuit.

V.3 The CCITT standard which defines the International Alphabet 5 code.

V.4 The CCITT standard which defines the International Alphabet 5 signal structure for transmission on public telephone networks.

V.6 The CCITT standard for modulation rates and data signaling rates for synchronous transmission on leased lines.

V.10 The CCITT standard for the electrical requirements for unbalanced double-current interchange circuits for use with integrated circuits.

V.11 The CCITT standard for balanced double-current interchange circuits for use with integrated circuits.

V.13 The CCITT standard which provides for simulated half-duplex (switched carrier) control. V.32 and V.33 compliant modems that support V.13 can be used in synchronous IBM RJE environments.

V.15 The CCITT standard for acoustic couplers.

V.16 The CCITT standard for special modems used in the medical field.

V.17 The CCITT standard for 14.4 Kbps half-duplex facsimile transmission.

V.19 The CCITT standard for parallel data transmission for modems over the public telephone network.

V.20 The CCITT standard for parallel data transmission for universal modems over the public telephone network.

V.21 The CCITT standard for 300 baud full duplex modems over public telephone networks.

V.22 The CCITT standard for 1200 baud full duplex modems over 2-wire leased or public telephone networks.

V.22bis The CCITT standard for 2400 baud full duplex modems over 2-wire leased or public telephone networks.

V.23 The CCITT standard for 600 baud and 1200 baud use over telephone networks.

V.24 The CCITT standard list of definitions for interchange circuits between data terminal equipment and data circuit terminating equipment.

V.25 The CCITT standard for automatic calling and/or answering equipment over telephone networks including the disabling of echo suppressers on manually established calls.

V.26 The CCITT standard for 2400 baud modems over 4-wired leased circuits.

V.26bis The CCITT standard for 1200/2400 baud modem for use in general switched telephone networks.

V.27 The CCITT standard for 4800 baud modems with manual equalizers for use on leased telephone type circuits.

V.27bis The CCITT standard for 2400/4800 baud modems with automatic equalizers for use on leased telephone circuits.

V.27ter The CCITT standard for 2400/4800 baud modems for use in the general switched telephone network.

V.28 The CCITT standard for the electrical characteristics for unbalanced double current interchange circuits.

V.29 The CCITT standard for 9600 baud modems for use in point-to-point leased line telephone circuits.

V.31 The CCITT standard for the electrical characteristics for single current interchange circuits controlled by physical contacts.

V.32 The CCITT standard for 9600 baud two wire full duplex modems for the general switched telephone network. Trellis-encoding modulation allows high data speeds and reduces errors.

V.32bis The CCITT standard for full duplex modems for the general switched telephone network on 2-wire lines at the following data rates: 14,400 bps, 12,000 bps, 9600 bps, 7200 bps, 4800 bps. It offers two advantages over V.32 -- first, data can be transmitted faster (up to 14,400 bps); and it redefines modem-connection negotiations (called training and retraining).

V.33 The CCITT standard for synchronous data transmission with full duplex operation over 4-wire leased lines with data rates at 14,400 bps or 12,000 bps. A V.33 compliant modem uses the same signal modulation techniques that are used by V.32 modems but one is restricted to using 4-wire leased lines.

V.35 The CCITT standard for data transmission at 48 Kbps using 60-108 KHz group-band circuits. It is typically used for DTEs or DCEs that interface to a high-speed digital carrier such as the AT&T Dataphone Digital Service (DDS).

V.36 The CCITT standard for synchronous transmission via modems over circuits operating at frequencies from 60 to 108 Khz.

V.40 The CCITT standard for error indications using electromechanical equipment.

V.41 The CCITT standard for a code independent error control system.

V.42 The CCITT standard for error control; it contains two algorithms (LAPM or Link Access Protocol, and MNP 1-4). When two V.42 compliant modems establish a connection they use LAPM to control data errors and retransmit bad data blocks. If one modem supports V.42 and the other supports only MNP, then the two negotiate to use the MNP protocol. In both cases, the error-control process is automatic and requires no special user actions or software programs. V.42 roughly corresponds to MNP Level 5. The difference is the amount of data compressed. V.42 can generate a 4:1 ratio of data compression depending on the file type. *see also* Microcom Networking Protocol.

V.42bis The CCITT standard to specify data compression techniques for modems which use LAPM error control in V.42 devices.

V.50 The CCITT standard for quality of data transmission circuits.

V.51 The CCITT standard for maintenance of telephone circuits used for data transmission.

V.52 The CCITT standard for equipment used to measure distortion and error rates in data and telephone circuits.

V.53 The CCITT standard on limits for the maintenance of telephone circuits used for data.

V.54 The CCITT definition of devices used doing loop back tests on modems.

V.55 The CCITT standard for equipment for measuring impulsive noise on telephone lines.

V.56 The CCITT standard for comparative tests for modems.

V.57 The CCITT definition of a data test set for measuring high data signaling rates.

VALUE ADDED NETWORKS (VAN) These are usually developed by companies which lease common carrier phone lines and develop different tariff structures. VANs provide services which build on the basic data transfer capabilities of the network to give subscribers additional benefits from the network for an fee. Common capabilities include access to commercial information utilities, electronic mail and airline tables. Examples in the United States would be Telenet and Tymnet.

VAX FTAM In DECnet Phase V, this file transfer, access and management protocol allows file management and transfers to OSI compliant systems.

VAX OSI TRANSFER SERVICE *see* VOTS.

VC *see* Virtual Circuit.

VC *see* Virtual Circuit Manager.

VERnet The Virginia Education and Research Network is a mid-level regional network of the NSFNET for the state of Virginia.

VERONICA Provides keyword indexing and access for menu offerings to different information systems on the Internet. Once appropriate information systems are identified, users may telnet to those systems to query databases and services of interest.

VERTICAL REDUNDANCY CHECK (VRC) An error checking routine that uses a parity bit for each character so that the total number of 1 bits in the character is odd or even as determined by the protocol.

VHF An abbreviation for "very high frequency."

VIDEO TELECONFERENCE A conference in which full-motion video is transmitted as well as voice and maybe graphics. The video signal can be one way (from one to many points) or two-way (simultaneously connecting two or more sites).

VIDEOTEX A graphics and textual based information retrieval system where data are usually transmitted over the telephone lines to home computers or terminals.

VIEWDATA An interactive information system over telephone lines in which users can access a host computer via a special terminal or microcomputer.

VINES VIrtual NEtworking System. A proprietary network operating system developed by Banyan Systems, Inc.

VIRTUAL Something that appears to exist but does not.

VIRTUAL CALL *see* Virtual Circuit.

VIRTUAL CIRCUIT A communications link that gives the user the impression that a dedicated line exists between the remote device and the host. This impression is due to the fact that data are processed in the order in which it was sent (just as with a real circuit). Contrast Datagram.

VIRTUAL CIRCUIT (CALL) MANAGER (VCM) A program which monitors and manages virtual circuit connections on an X.25 network.

VIRTUAL LIBRARY The act of remote access to the contents and services of libraries and other information services, combining an on-site working collection in all formats with an electronic network which provides access to and delivery from external libraries and commercial information sources. Knowledge and information may be accessed on a worldwide basis.

VIRTUAL PRIVATE NETWORK (VPN) A network which offers the functionality of a dedicated private network using the facilities of a public switched network.

VIRTUAL TERMINAL (VT) A remote login terminal emulation utility program for OSI.

VIRTUAL TERMINAL SERVICE (VTS) *see* Virtual Terminal.

VISTANET One of five gigabit network research testbeds established by NREN. Located in North Carolina, this network explores remote medical imagery and diagnosis.

VOICE MAIL The ability to store and retrieve voice messages from the public telephone system.

VOTS VAX OSI Transport Service. Provides OSI transport layer functions in DECnet Phase V and supports interworking among different subnetworks by using the Internet Protocol, which provides Connectionless Network Services over local area networks and X.25 wide area networks.

VPN *see* Virtual Private Network.

VRC *see* Vertical Redundancy Check.

VT *see* Virtual Terminal.

VTS An abbreviation for "virtual terminal service." *see* Virtual Terminal.

VU/TEXT INFORMATION SERVICES A major bibliographic utility which has full text newspapers. It is a subsidiary of Knight Ridder.

W3 *see* World Wide Web.

WAIS Wide Area Information Server. The WAIS project began as an experimental venture between Thinking Machines Corporation, Apple Computer, Dow Jones & Company and KPMG Peat Marwick. The purpose was to create an easy-to-use interface which could access many information servers on the Internet, regardless of location. WAIS uses client/server architecture based on an extension of the Z39.50 NISO protocol. Two components are necessary, WAIS servers and WAIS client software. Interaction with the WAIS system occurs through the Question Interface which is a graphical user interface piece of client software. To begin a session the user asks a question, pulls down a menu identifying servers which will be queried (the WAIS software can also identify source systems if the user does not know what to select), and then sends the query. After the appropriate information is retrieved from the remote servers, headlines of materials are displayed in a window and the user may select the relevant information.

WAN Wide Area Network. A communications network that spans large areas (hundreds or thousands of miles) by using telecommunications lines provided by a common carrier (e.g. the phone company).

WATS Wide Area Telecommunications Service. Allows phone customers to make long-distance calls and have them billed on a bulk basis rather than individually.

WAVE DIVISION MULTIPLEXING Transmitting two or more separate channels of data over an optical fiber using different wavelengths of light.

WAVEGUIDES These microwave pipes can be used to transmit microwave signals and are often used to transmit microwave signals up and down microwave antenna towers.

WESTERN LIBRARY NETWORK *see* WLN.

WESTLAW A major bibliographic utility with full text legal information operated by West Publishing Company.

WESTNET The mountain states (U.S.) network connected to the NSFNET. An academic and industrial network which connects many smaller regional networks such as the New Mexico Technet and the Colorado Supernet. It serves Utah, Arizona, Colorado, New Mexico and Wyoming.

WHITE NOISE Background or random acoustical or electrical signals.

WHITE PAGES Lists or directories of users that are accessible through the Internet.

WHOIS SERVERS An interactive software program that allows network users to register their names and electronic mail addresses and to search for this type of information on other registered users. The QUIPU WHOIS is a well known implementation of this program run by the Defense Data Network (DDN) and can be accessed by Telneting to nic.ddn.mil and running the WHOIS program.

WIDE AREA INFORMATION SERVER *see* WAIS.

WIDE AREA NETWORK *see* WAN.

WIDE AREA TELECOMMUNICATIONS SERVICE *see* WATS.

WILSONLINE A major bibliographic utility operated by H.W. Wilson Company.

WIN *see* DFN.

WINDOW A mechanism that allows one or more packets of data to be sent without having to wait for a response from the receiving device.

WIRELESS LANS A technique for connecting workstations in local area networks in which data are transmitted via packet radio, laser signals or infra-red signals.

WISCNET The Wisconsin Network supports high-speed networking for academic and government institutions in the state of Wisconsin. It provides no direct services to end-users but provides Internet access to the member institutions. It is connected to the CICnet regional network.

WLN Western Library Network. A bibliographic information utility which supports libraries in shared cataloging, retrospective conversion and other activities.

WORKSTATION A device, often a microcomputer, which serves as an interface between a user and a file server or host computer.

WORLD WIDE WEB (WWW or W3) World Wide Web. An Internet navigational tool initially developed at CERN, the European Particle Physics Laboratory in Geneva, Switzerland. It is an effort to organize information on the Internet plus local information into a set of hypertext documents. A person navigates the network by moving from one document to another via a set of hypertext links. The WWW may be used via a "line oriented" front-end called a "browser" or one may load a WWW client for a graphical user interface.

WORM Write once read many. An acronym for optical disc technology in which data can be written once but the data cannot be erased.

WVNET West Virginia Network for Educational Telecomputing connected to the NSFNET.

WWW *see* World Wide Web.

WXMODEM A file transfer protocol developed by Peter Boswell which is a variation on XMODEM. It offers CRC error-checking and sliding windows (not provided in XMODEM). It is especially useful when transferring files over public data networks.

X SERIES The collection of CCITT standards for communications interfaces for data terminal equipment (DTE) and data circuit terminating equipment (DCE) (e.g. network interface devices such as modems).

X.1 The CCITT standard which defines classes of service for international public data network users.

X.2 The CCITT standard which defines international user facilities in a public data network.

X.3 The CCITT standard for Packet Assembler/Disassembler (PAD) parameters. It specifies how character strings coming from a communications device are to be converted to X.25 packets and vice versa.

X.4 The CCITT standard which defines the structure of signals to the International Alphabet 5 over a public data network.

X.12 *see* Electronic Data Interchange.

X.20 The CCITT standard which defines the start/stop interface for DTE and DCE devices over public data networks.

X.20bis The CCITT standard which defines the start/stop interface for DTE and DCE devices with asynchronous V-series modems over public data networks.

X.21 The CCITT standard which defines the DTE and DCE interface on synchronous devices over public data networks.

X.21bis The CCITT standard which defines the DTE and DCE interface on synchronous V-series modems over public data networks.

X.24 The CCITT definition for interchange circuits between DCE and DTE devices over public data networks.

X.25 The CCITT standard that defines a protocol for gaining access to public packet switching networks. X.25 uses packet switching to provide virtual circuits between nodes in a network.

X.28 The CCITT standard that describes the interface protocol for data terminal equipment to a PAD connection.

X.29 The CCITT standard that prescribes the protocol for a PAD to terminal connection.

X.31 The CCITT standard which provides how data terminal equipment can be connected to an ISDN network.

X.32 The CCITT standard which describes how X.25 data terminal equipment can connect to a packet switched X.25 network using switched access lines.

X.75 The CCITT standard which defines the message structure required for gateway nodes of public packet switching networks. X.75 uses virtual circuits to establish connections between dissimilar networks. The standard indicates how packets are composed and the procedures for sending and receiving them.

X.121 The CCITT standard for numbering networks and stations which are interconnected using X.25 protocols.

X.223 The CCITT standard for using the X.25 protocol to provide the OSI connection-mode network service.

X.400 The CCITT standard for electronic mail used in OSI. In OSI this would be an application utility. The ISO name is Message-Oriented Text Interchange System (MOTIS).

X.500 The CCITT standard for the OSI Directory Service (its counterpart is the ISO/IEC 9594 standard). The directory is a database which contains information such as names of people, applications, and devices on the network. It includes structured mechanisms for accessing the service either by individuals or through an OSI application. It uses a hierarchical naming system similar to the Domain Name Service (DNS) used in TCP/IP networks.

X.610 The CCITT standard which defines the provision and support of the OSI Connection-Mode Network Service.

X.613 The CCITT standard which defines the procedures for the OSI Connection-Oriented Network Service by packet-mode terminal adapters using the ISDN.

XENIX A version of UNIX developed for the IBM AT class of microcomputers.

XFERIT A Macintosh application which allows users of MacTCP to transfer files to and from FTP archives.

XI IBM X.25 SNA Interconnect. A software program developed by IBM to support 3225 node processors to support X.25 equipment (DTEs) to communicate over SNA networks.

XID Exchange Information. A command and response used in the HDLC protocol to verify the identity of the called station by the caller.

XMODEM A public domain 8-bit error checking protocol developed for file transfers between computers. It was developed in the 1970s by Ward Christensen and uses a 128 byte data block and CRC or checksum error checking routines.

XNS Xerox Network Systems. A local area network architecture developed by Xerox Corporation but never became popular as a communications standard. However, proprietary variations of this architecture are widely used.

X-ON/X-OFF A set of control characters which control (i.e. start and stop) the flow of data from a computer to an I/O device or terminal.

X.PC A proprietary network interface specification by Tymnet which provides the protocol for accessing their X.25 packet switching network using a switched access line.

XPDU An abbreviation for "X Protocol Data Unit."

XWINDOWS A distributed, device independent, network-transparent, multitasking windowing and graphics system used heavily in the UNIX environment.

YMODEM BATCH A file transfer protocol which is an improved version of XMODEM but offers 1024-byte block size, batch transfers, and includes the filename and file size for the transferred files.

YMODEM-G BATCH A file transfer protocol similar to the 1K-XMODEM-G protocol except that it offers no error checking since it was designed to work with an error-checking modem. It is very efficient when used with the proper equipment.

ZMODEM A file transfer protocol developed by Chuck Forsberg and is very efficient due to 1024-byte blocks, allows batch transfers, automatically sends the file's name, size and date attributes, and offers sliding windows.

Z39.50 1988 A NISO standard entitled "Information Retrieval Service Definition and Protocol Specification for Library Applications. This standard offers a protocol that provides for the exchange of messages between computers for the purpose of information retrieval. It has important applications for library and information service vendors and it gives guidelines for the format of queries, provides for the transfer of database records, and defines other record types. This standard is designed as an application layer (layer 7) protocol within the Open Systems Interconnection (OSI) protocol suite but is also being mapped into the TCP/IP protocol suite. The ISO equivalent standard is usually referred to as "ISO Search and Retrieve" (SR). *see also* Linked Systems Protocol.

Z39.58 *see* Common Command Language.

References

Adams, Roy. *Communication and Delivery Systems for Librarians.* Brookfield, VT: Gower Publishing Company, 1990.

Avram, Henriette D. et al. "Networking in Transition: Current and Future Issues." *Library Hi Tech* 6(4):101-119 (1988).

Bailey, Charles W. et al. *The Public-Access Computer Systems Review, Volume 1, 1990.* Chicago, IL: Library and Information Technology Association, 1992.

Bell, Gordon. "Steps Toward a National Research Telecommunications Network." *Library Hi Tech* 6(1):33-36 (1988).

Berners-Lee, Tim et al. "World Wide Web: The Information Universe." *Electronic Networking* 2(1):52+ (Spring 1992).

———. "The World-Wide Web." *Computer Networks and ISDN Systems* 25(4/5):454+ (November 1, 1992).

Black, Uyless. *Data Networks.* Englewood Cliffs, NJ: Prentice Hall, 1989.

Boss, Richard W. *Telecommunications for Library Management.* White Plains, NY: Knowledge Industry Publications, 1985.

Brown, Charles D. "IEEE Standards and Issues." *Information Technology and Libraries* 9(1):89-93 (March 1990).

Clarkson, Mark. "The Many Flavors of SQL." *Byte* 18(7):109-112 (June 1993).

Cole, Gerald D. *Implementing OSI Networks.* New York, NY: John Wiley & Sons, 1990.

Collier, Mel, editor. *Telecommunications for Information Management and Transfer: Proceedings of the First International Conference held at Leicester Polytechnic, April 1987.* Brookfield, VT: Gower Publishing Company, 1987.

Cronk, Randall D. "EISes Mine Your Data: Executive Information Systems and Client/Server Computing Help You Understand Your Data Better." *Byte* 18(7):121-128 (June 1993).

Datapro Reports on Telecommunications. Delran, NJ: McGraw-Hill Incorporated, 1990.

Deering, Stephen E. "SIP: Simple Internet Protocol." *IEEE Network* 7(3):16+ (May 1, 1993).

DeLoughry, Thomas J. "Software Designed to Offer Internet Users Easy Access to Documents and Graphics." *The Chronicle of Higher Education* 39(14):A23 (July 7, 1993).

Derfler, Frank J., Jr. "Networking Acronyms & Buzzwords (Guide to Networking Terminology)." *PC Magazine* 7(11):99-105 (June 14, 1988).

Dern, Daniel P. "Applying the Internet." *Byte* 17(2):111+ (February 1992).

Deutsch, Peter. "Resource Discovery in an Internet Environment The Archie Approach." *Electronic Networking* 2(1):45+ (Spring 1992).

Eadie, Gavin R. "Migrating from Centralized to Distributed E-Mail." *Query Higher Education,* Number 9:2-5 (Spring 1993).

Engle, Mary E. et al. Internet Connections: A Librarian's Guide to Dial-up Access and Use. Chicago, IL: *Library and Information Technology Association,* 1993.

Fenly, Judith G., and Beacher Wiggins, editor. *The Linked Systems Project: A Networking Tool for Libraries*. Dublin, OH: OCLC Online Computer Library Center, 1988.

Fisher, Sharon. "Getting Away from Cables: Wireless LANs Let PCs Communicate without Cabling." *Info World* 12(18):S1 (April 30, 1990).

"Glossary of LAN Terms and Acronyms." *Info World* 11(29):52 (July 17, 1989)

Gordon, M., A. Singleton, and C. Rickards. *Dictionary of New Information Technology Acronyms*. Detroit: Gale Research Company, 1984.

Graham, John, and Sue J. Lowe. *The Facts on File Dictionary of Telecommunications*. Revised Edition. New York: Facts on File, 1991.

Graves, Randall, and Russell Clement. "Telecommunications: A Primer for Librarians." *Wilson Library Bulletin* 63(5):50+ (January 1, 1989).

Helmers, Scott A. *Data Communications*. Englewood Cliffs, NJ: Prentice Hall, 1989.

Heterick, Robert C. Jr. "Networked Information: What Can We Expect and When?" *CAUSE/EFFECT* 13(2):9-14 (Summer 1990).

Hewitt, Joe A., editor. *Advances in Library Automation and Networking: A Researching Annual*. Greenwich, CN: JAI, 1987.

Interlan on Interoperability. Boxborough, MA: Interlan, Inc., 1988.

Jacob, M.E.L., editor. *Telecommunications Networks: Issues and Trends*. White Plains, NY: Knowledge Industry Publications Inc., 1986.

Kahle, Brewster, and Art Medlar. "An Information System for Corporate Users: Wide Area Information Servers." *Online* 15(5):56-60 (September 1991).

Kemper, Marlyn. *Networking: Choosing a LAN Path to Interconnection*. Metuchen, NJ: The Scarecrow Press, 1987.

Kesselman, Martin. "Beyond Bitnet: Telnetting to the United Kingdom." *College & Research Libraries News* 54(3):134-136 (March 1993).

King, Donald W. et al. *Telecommunications and Libraries: A Primer for Librarians and Information Managers*. White Plains, NY: Knowledge Industry Publications, 1981.

Krause, Joanne. *Procomm Plus 2.0 at Work*. Reading, MA: Addison-Wesley Publishing Company, Inc., 1991.

Krol, Ed. "Internet Users Building Electronic Card Catalog." *Network World* 10(20):33 (May 17, 1993).

———. *The Whole Internet User's Guide & Catalog*. Sebastapol, CA: O'Reilly & Associates, 1992.

"LAN Links Take Form of Repeaters, Bridges, Routers and Gateways." *PC Week* 6(4): C23 (January 30, 1989)

Lane, Elizabeth, and Craig Summerhill. *Internet Primer for Information Professionals: A Basic Guide to Internet Networking Technology*. Westport, CT: Meckler Corporation, 1993.

Learn, Larry L. "Networks: A Review of Their Architecture and Implementation." *Library Hi Tech* 6(2):19-49 (1988).

———. "Networks: The Telecommunications Infrastructure and Impacts of Change." *Library Hi Tech* 6(1)13-31 (1988).

———. *Telecommunications for Information Specialists.* Dublin, OH: OCLC Online Computer Library Center, 1989.

Lynch, Clifford A. "Access Technology for Network Information Resources." *CAUSE/EFFECT* 13(2):15-20 (Summer 1990).

———. "From Telecommunications to Networking: The MELVYL Online Union Catalog and the Development of Intercampus Networks at the University of California." *Library Hi Tech* 7(2):61-83 (1989)

———. "Library Automation and the National Research Network." *Educom Review* 24(3):21+ (Fall 1989).

———. "Linking Library Automation Systems in the Internet: Functional Requirements, Planning, and Policy Issues." *Library Hi Tech* 7(4):7-18 (1989).

Machovec, George S. "Brief Glossary of Networking and Telecommunications Part 1." *Online Libraries and Microcomputers* 7(4):1-7 (April 1989).

———. "Brief Glossary of Networking and Telecommunications Part 2." *Online Libraries and Microcomputers* 7(5):1-8 (May 1989).

———. "Brief Internet and NREN Glossary: Part I (A-L)." *Online Libraries and Microcomputers* 11(5):1-4 (May 1993).

———. "Brief Internet and NREN Glossary: Part II (M-Z)." *Online Libraries and Microcomputers* 11(6):1-4 (June 1993).

———. "Resources on the Internet: Archie and Gopher." *Online Libraries and Microcomputers* 10(10):1-4 (October 1992).

———. "TCP/IP and OSI: Networking Dissimilar Systems Implications for Libraries." *Online Libraries and Microcomputers* 7(6-7):1-4 (June 1989)

Moholt, Pat. "The Influence of Technology on Networking." *Special Libraries* 80(2):113+ (Spring 1989).

Myers, Robert A. *Encyclopedia of Telecommunications.* San Diego: Academic Press, 1989.

Nauman, B. Clifford. "Prospero: A Tool for Organizing Internet Resources." *Electronic Networking: Research, Applications and Policy* 2(1):30-37 (Spring 1992).

Nickerson, Gord. "The Internet Gopher." *Computers in Libraries* 12(8):53-56 (September 1992).

Page, Mary. "A Personal View of the Internet." *College & Research Libraries News* 54(3):127-132 (March 1993).

Perry, Dennis G., Steven H. Blumenthal, and Robert M. Hinden. "The ARPANET and the DARPA Internet." *Library Hi Tech* 6(2):51-62 (1988).

Polly, Jean Armour. "Surfing the Internet." *Wilson Library Bulletin* 66(10):38+ (June 1, 1992).

Scott, Peter. "HYTELNET as Software for Accessing the Internet: A Personal Perspective on the Development of HYTELNET." *Electronic Networking: Research, Applications and Policy* 2(1):38-44 (Spring 1992).

Shaw, Dennis. "Library Networking in Europe." *IATUL Quarterly* 3(2):74-81 (June 1989).

Sherron, Gene T. "ISDN Take Another Look." *CAUSE/EFFECT* 13(2):3-5 (Summer 1990).

———. *An Information Technology Manager's Guide to Campus Phone Operations*. CAUSE Professional Paper Series, #3. Boulder, CO: CAUSE, 1990.

Sloane, Bernard G., and J. David Stewart. "ILLINET Online: Enhancing and Expanding Access to Library Resources in Illinois." *Library Hi Tech* 6(3):95-101 (1988).

Smith, Christine H., editor. *Open Systems Interconnection: The Communications Technology of the 1990's: Papers from the Pre-Conference Seminar Held at London, August 12-14, 1987*. IFLA Publications 44. New York: K.G. Saur, 1988.

Stein, Richard Marlon. "Browsing Through Terabytes." *Byte* 16(5):157+ (May 1991).

Thompson, M. Keith, and Kimberly Maxwell. "Connectivity: Building Workgroup Solutions: Networking CD-ROMS." *PC Magazine* 9(4):237+ (February 27, 1990).

Updegrove, Daniel A., John A. Muffo, and John A. Dunn Jr. "Electronic Mail and Networks: New Tools for University Administrators." *Educom Review* 25(1):21-28 (Spring 1990).

White, Gene. *Internetworking and Addressing. Uyless Black Series on Computer Communications.* New York: McGraw-Hill, 1992.

Wilson, David L. "Array of New Tools is Designed to Make it Easier to Find and Retreive Information on the Internet." *The Chronicle of Higher Education* 39(38):A17-A19 (May 26, 1993).

George S. Machovec is the Technical Coordinator for the Colorado Alliance of Research Libraries (CARL) in Denver, Colorado. He has been with CARL since January 1993 and is responsible for managing the technical affairs of the nonprofit Alliance. Prior to that he was the Head, Library Technology and Systems at Arizona State University (Tempe) where he was responsible for the integrated library system and other aspects of information technology. Previous to that he held several other positions at ASU including head of the Solar Energy Collection and coordinator of computer reference service. In 1977, Machovec received his MLS from the University of Arizona (Tucson).

Machovec is the managing editor for *Online Libraries and Microcomputers* (1983-) and is the software review editor for LITA's *Information Technology and Libraries* (1990-). He has also been a member of LITA's Publication committee and has been involved with ALA since 1977. Machovec has been widely published in library automation and information technology literature including books, journal articles and conference papers.